THEOLOGY AND BIOTECHNOLOGY

Implications for a New Science

CELIA DEANE-DRUMMOND

London and Washington

Geoffrey Chapman
A Cassell imprint
Wellington House, 125 Strand, London WC2R 0BB
PO Box 605, Herndon, VA 20172 USA

First published 1997

British Library Cataloguing-in-Publication Data
A catalogue record for this book is available from the British Library.

ISBN 0–225–66851–3 (Hardback)
0–225–66849–1 (Paperback)

Typeset by Keystroke, Jacaranda Lodge, Wolverhampton
Printed and bound in Great Britain by Biddles Ltd, Guildford and King's Lynn

Contents

Acknowledgements

I would like to thank all those who have indirectly helped to shape this book through numerous discussions, both formal and informal. In particular, I would like to thank colleagues at the International Consultancy on Religion, Education and Culture (ICOREC), in which I was actively involved from 1991 to 1994 and where some of the ideas for this book were formed. I would like to thank Andrew Dawson and David Major for helpful suggestions and comments for Chapters 5 and 6 respectively. I would also like to thank Professor Richard Bauckham, who supervised my PhD thesis on Jürgen Moltmann's doctrine of creation, which I completed in 1992, aspects of which are reproduced here in a modified form. I have acknowledged those journals in which some of the ideas appear in the relevant chapters. I would like to thank Robin Grove-White and Bron Szerszynski, whom I visited in June 1996 at the Centre for the Study of Environmental Change (CSEC) in Lancaster, who gave me some very valuable insights and stimulus. I would like to acknowledge the support of Michael Walsh, who, as editor for Geoffrey Chapman, spurred me on to complete this book by his continued support. Finally, many thanks to Leilah Nadir, editor for Cassell, for her helpful advice and comments.

Introduction

The idea for this book first occurred to me over ten years ago when I was a lecturer in plant physiology. As a committed Christian I was concerned to see how my faith made sense in the light of my scientific work. I became aware of the quite revolutionary changes taking place in botany departments. Those that were considered by the government and grant awarding bodies to be the most forward thinking were those who linked their research to particular applications in the newly emerging field of biotechnology. However, it was not simply an application of known theory for commercial use, but a reorientation of the aims of academic research so that it addressed issues that were seen to be of commercial interest. This inevitably entailed a specialization into the fields of molecular biology and genetic engineering of plants. Those at the other end of the spectrum, namely the ecologists, felt themselves to be marginalized by the increasing strength of the molecular biology lobby. In fact the natural historians were considered to be 'out of touch' with the exciting and rapid advances made in this new biology. As a plant physiologist I was most concerned with the way plants function, using the tools of biochemistry and biophysics. Yet in order to understand the way plants work as a whole, you have to consider the environment. Hence, I was caught somewhere in the middle of the battle for supremacy between the ecologists and the biotechnologists. Furthermore, I believed at this time that not enough was being done to encourage a more holistic approach to science which looked at the overall social consequences of research policy.

Since this time there has been extensive theological reflection on environmental issues. However, the questions raised by biotechnology seem to have been ignored or restricted to narrow issues relating to the genetic engineering of humans. This book is about the implications of biotechnology in the non-human sphere, though I concentrate more on key examples from agricultural botany. The current BSE crisis, the anxiety about genetically engineered tomato and soya now appearing on supermarket shelves, all these are trends which are just the tip of the iceberg. Yet the academic research was fostered and spread a decade ago. Why has this happened? I hope to begin to answer this question in these preliminary remarks. I also hope to show how theology can contribute to this debate and argue for a new way of approaching science that is more aware of its social and ethical consequences.

Biotechnology relies on a view of the world which treats 'nature' as other than human and all living things are viewed as a mechanism. Such a belief is born of the mechanistic philosophy stemming from the ideas of Descartes. Those who reacted against this philosophy in the eighteenth century developed a school of thought known as 'nature philosophy'.[1] In physics there was a similar reaction against the static Newtonian ideas, though in this case the new physics was indebted to Einstein. The greater degree of uncertainty which is part of Einstein's thesis allowed a more mystical view of physics to emerge, such as the well known *The Tao of Physics*.[2] The difference between the new physics and the new biology is that while the former seemed to challenge the idea of mechanism, the latter still relies on a Newtonian understanding of physics. James Lovelock has tried to challenge the mechanistic approach to living things through his Gaia hypothesis.[3] However, his ideas about interconnectedness are still on the fringe, rather than in the mainstream, of biology. His theory encompasses geology, earth science, biology in a complex way which by its nature is very hard to 'prove' in an acceptable manner to a mind trained in scientific method. By contrast, Einstein's ideas remain testable and so are considered to be valid as science, even though they leave space for discussion of other dimensions not normally accepted as the province of science.

The challenge for Christian theology is how far any of these spiritualities can reformulate science in a way that shows an ethical responsibility. Furthermore, I suspect that those who are engaged in biotechnology have little time for broad, seemingly untestable hypotheses such as the Gaia hypothesis, even though its apparent religious resonance has an appeal for those interested in religious ideas. Should biotechnology just be left to its own devices and allowed to influence our culture in ways which may not be acceptable? I hope that this book will be read not just by Christians, but by anyone interested in the introduction of moral values into the practice of science and technology. If market forces are driving the mushrooming of biotechnology, then public opinion can help to shape future directives in a science that is reliant on commercial interest and funding. Furthermore, those in positions of responsibility in policy making can decide where the boundaries need to be drawn. The old scientific adage that anything that is possible is worthwhile simply does not hold water any more.

One of the difficulties facing scientists is that their training is such that only that which is scientifically testable is deemed to be valid. In other words, unless it can be proved by experiment it counts as nothing and is mere opinion. The difficulty with this mindset is that any theological reflection is likely to be discounted. Richard Dawkins, for example, rejects theological writing because it is never 'testable by any standards of evidence: standards that might be respected by scientists or by lawyers or by historians or by common sense'.[4] He believes that putting a God hypothesis against scientific reasoning is exposed as having no more 'solid basis than fairies'. Furthermore, the culture of the time views religious opinion in particular as matters for private, rather than public, discussion. The assumption in this view is that only certain ways of thinking count as knowledge, that is experimental ways of thinking and testing. Scientifically proven data seemingly provide hard objective knowledge that goes beyond the opinion of the observer.

I would argue against this view in two respects. First, since data can be used to support political ideas, the concept that knowledge is purely objective, just because it relies on scientific method, is naive. Consider, for example, the way the slave trade was justified by the seeming scientific evidence that humans with dark skin were further back on the evolutionary path towards progress and closer to the 'ape'. More recently this trend has continued in an alarming way in the 'evidence' that the IQ of black children is lower than that of white children.[5] Science is, of necessity, bound to be influenced by the particular views and assumptions of the scientists, either acknowledged or hidden as part of the cultural matrix. It is false to suggest that science is value-free. Secondly, theology does have something to say when it comes to issues of the survival of humanity and justice in the human community. Theology encourages a broader perspective which is hard to generate from science alone. This does not mean theology competes in the matter of particular areas of knowledge. The battle between the creationists and the evolutionists seems to me to be a 'strange contest'. To suggest that theology should try and ape science is rather like asking the question 'Should an elephant be like a concrete mixer?'[6] Strange because the Bible was never meant to be written as a scientific document. Strange because writers such as Dawkins can never resist a controversy and, as I hope to show below, he conveys in his writings anthropomorphisms that certainly go way beyond the scientific evidence alone. Furthermore, why have we come to regard scientific knowledge as the only way of knowing that is valid? Mary Midgley puts the point succinctly:

> The old wide notion of knowledge or learning as an aspect of wisdom – an element in the whole spiritual life which the West once shared with other cultures, has gradually shifted and narrowed to crystallise in the ideal of science. This too has been narrowed, so that it is now often conceived as simply exact information about the physical world, acquired professionally by experts using experimental methods.[7]

Yet certain scientists, possibly encouraged by the media and popular interest, seem to be convinced that the battle has to go on. The public status of science, raised even further by its practical usefulness, ensures an expectant audience. Richard Dawkins, for example, has carried out an almost evangelical campaign against religious believers. For him, all living organisms seem to be like robots dictated to by the commands of their selfish genes.[8] Human beings are like 'survival machines' for the genes they contain. Although, as a scientist, he claims that he cannot acknowledge purpose in the universe, he has convinced himself that he is justified in using purpose in a metaphorical sense. For him, the purpose of genes is to survive to the next generation, a gene 'leaps from body to body down the generations, manipulating body after body in its own way for its own ends, abandoning a succession of mortal bodies before they sink in senility and death'.[9] Now I would challenge strongly his assumption that the language of purpose that he uses remains metaphorical. Rather, he seems to be putting a degree of power into genes in a way that smacks of animism, which he would no doubt rigorously deny himself.[10] It is the seeming power of human understanding of genetics that has seduced Dawkins into an almost religious commitment to the genes as agents of immortality. It is this underlying philosophy which fosters the seduction of biotechnology, drawing as it does on the potential for transference of power through manipulation of genetic material. His official scepticism of any 'religious' ideas covers up an unwarranted dogmatism. Its apparent clarity of method feels more secure than the vaguer ideas associated with theological reflection. The hope in science becomes extended to analysis of human societies as such through sociobiology, bringing with it a more sinister threat of manipulation through 'genetic window-shopping'.

This trend in biology itself, while it has been designed to reject theism of Christianity, seems to me to encourage a more holistic approach which draws on traditional understandings of the relationships between God, humanity and the environment. The logical conclusions of Dawkins's arguments, namely that humans are mere machines, strikes the listener as incredible and certainly less credible than traditional Christianity. The isolation of genetics and biotechnology produces a lopsided view of the world and humanity which leads to further distortions in environmental and social terms. Charles Darwin's thesis on the theory of evolution was only possible in the context of evolutionary ideas about society, such as that of Herbert Spencer.[11] He was reluctant to publish because he knew it would create a storm in the church, seemly weakening the status of humans to be that of animals. Richard Dawkins, similarly, is only able to propagate his views in the context of a cultural affirmation of biotechnology and molecular biology. Yet, unlike Darwin, he is only too keen to dismiss all religious claims. Now it seems to me that his views do not just weaken the place of humanity in relation to animals, but all living things become agents of the 'replicators'. Life is reduced to the lowest common denominator, the gene, which alone survives to the next generation. I very much doubt if Darwin would have recognized his idea of the 'survival of the fittest' in Dawkins's concept of the survival of the genes in isolation, even though Darwin held to the concept of generation.[12] This is because in Darwin's time there was still a strong belief in progress; that is, that evolution was working towards forms which were more advanced and complex. It is hard to find any idea of a real increase in complexity in Dawkins's views, other than that generated by the spontaneous interaction of genes with their 'host machine', rather like the programming of a computer. He insists that genes, too, for all intents and purposes, are basically random constructs arising spontaneously over millennia of evolutionary processes. It seems to me that it takes far more faith to believe this story that the traditional Christian version of creation by action of a purposeful Creator God.

What sort of boundaries should be set by those practising biotechnology? Some of these are discussed in a book which has just been published entitled *Improving Nature?: The Science and Ethics of Genetic Engineering.*[13] As its name implies this book seeks to show how science might add to the natural world and 'improve' what is there already. The authors explain the science of genetic engineering prior to a brief section devoted to moral, ethical and theological issues, followed by a more in-depth exploration of particular examples of plant, animal and human genetic engineering. In general, their arguments are weighted towards uncovering the facts of science for educational purposes and as a way of defending genetic engineering, which they believe has suffered an unfair press. The theological content of the book is necessarily rather superficial and mentions only briefly a range of religious perspectives.

The authors, Reiss and Straughan, distinguish between areas of 'moral concern', which arise out of a public sense of right or wrong, and areas of 'ethical concern', which are to do with the legitimacy or otherwise of particular actions.[14] They address the reasoning behind a consideration that genetic engineering is wrong in itself (intrinsic) or wrong because of its consequences (extrinsic). They believe that an assessment of risk has to be taken into account if genetic engineering is to be a responsible activity that can be justified. They believe that all our actions in scientific research carry some risk, the question is whether or not the level of risk is acceptable. They point out that all species have changed over time and some organisms, such as viruses, have developed the capacity to carry genetic material between species. They argue that these facts mean that it does not make sense to reject genetic engineering as 'unnatural'.[15] However, while I agree that species are not genetically isolated, it seems to me that the ability of genetic engineering to speed up the process to months or years and include many different species, especially those that are highly evolved, does make transgenetic manipulations in some sense 'unnatural'. As they acknowledge, most new species emerge over a period of more than a million years, with only a handful taking less than 5000 years. The incidence of 'natural' transgenetic occurrences is very rare.

Objections to genetic engineering have to go beyond a consideration of naturalness or otherwise of an event. In other words what is natural is not always 'good'. However, while this amounts to a weak philosophical argument against genetic engineering, it is equally fallacious to suggest that this somehow supports 'unnatural' interference. I have the impression that the authors want to defend genetic engineering against some of the more exaggerated reactions to the process. They argue, for example, that biologists are, if anything, more likely to respect the living material they are working with, compared with others. I am less optimistic about the motive of genetic engineers, given the commercial pressure that has come to bear on molecular biology research. However, I also believe that the issue is far wider than just scientists, who have become scapegoats. The will for change has to include the whole community. More disconcerting, perhaps, is that the official regulations in both the United States and Europe over the testing and use of micro-organisms in the field are becoming more relaxed, rather than more stringent.[16] I would agree with the authors' suggestion that for the genetic engineering of plants :

> a cautious, well regulated, step by step approach is the most responsible way forward. The socio-economic effects are even more difficult to predict, and there is a clear ethical obligation upon those countries and institutions that will profit from the new technology to minimise the losses of the most vulnerable economies and individuals.[17]

I begin this book in Chapter 1 with a discussion of issues in science and faith that are most relevant to a discussion of biotechnology in the context of environmental issues. In particular, I hope to show how theology is relevant to environmental questions. There has been a tendency in Christian theology to identify too closely with the natural world or to become alienated from it. I aim to show how bridges between science and spirituality were formulated in the history of science. Modern scientists presupposed a separation of the world from God, which is possible for those in the Judaeo-Christian tradition. Some writers have even gone as far as to claim that the biblical notion of 'dominion' of the earth encouraged 'domination' and subsequent exploitation. With the emergence of the modern scientific method and the theses of Charles Darwin, the idea of God no longer seemed relevant to a study of biology. The secularization of natural history, in turn, is believed by some theologians to be the root of the ecological crisis. Theologians and scientists, it seemed, retreated in the wake of fierce criticism of all kinds of 'natural theology', which sought to see the hand of God in the design of the natural world. The split between the two fields has led to mutual suspicion on both sides. The main locus of concern for theology now seemed to be on history, rather than 'nature'. More recently, theological reflection has moved towards a greater sensitivity to issues of creation, with a sharp rebuttal of the claim that Christianity has in some way been to blame for the ecological crisis. A more positive reappraisal is needed, with a return to the philosophy of Martin Heidegger, who sought to reformulate technology so that it becomes in tune with evolving natural systems. Furthermore, theology can contribute its own particular insights by use of metaphors which help to reshape our understanding of the relationships between God, humanity and our natural environment.

Chapter 2 explores different theologies of creation, both past and present. I believe that this is an important grounding for a discussion of the more specific issue of the relationship between theology and biotechnology. Indeed, this relationship needs to be grounded in a creation theology if it is to have any depth in theological terms. Issues in science and religion tend to be discussed in isolation from creation theology. In their turn, theologies of creation tend to be divorced from the real issues which emerge in scientific developments. The changing attitudes to the natural world in the early and early modern period put our current fascination with the natural world in perspective. I explore contemporary theologies from a range of perspectives, including that of Karl Barth, Teilhard de Chardin, Karl Rahner, Hans Küng, Eastern Orthodoxy, feminist theology and Jürgen Moltmann.

Environmental ethics form the subject matter of the third chapter. I explore basic issues in environmental ethics as a way of discerning the ethical and philosophical basis of value in the natural world. I address the question of the Gaia hypothesis in more detail. Can this hypothesis, which seems to marry religious and scientific ideas, offer ethical guidelines in how to treat our planet? I explore the question of animal rights and ask how far we can move from an ethic which is centred on human interest to one that is more concerned with animal welfare and the welfare of all living things, including plants. I offer the suggestion that a more inclusive theocentrism provides a positive way forward which avoids the negative rejection of the world in a narrow anthropocentrism or the seeming rejection of humanity in a biocentrism.

The fourth chapter examines specific cases of biotechnology, drawing on recent examples of genetic engineering. I hope to show how genetic engineering can be viewed as a promise, the idea that we somehow improve nature, which I discussed earlier in this introduction. However, genetic engineering also has a more sinister aspect and can become a means of powerful manipulation and a threat to the cultural and ecological stability on earth. While ecological systems are, in themselves, not static, in the course of evolution change usually happens gradually. Exceptions include the early massive 'pollution' of the earth by oxygen, which then led to completely new species emerging which relied on oxygen for growth.[18] The point I am making is that the interference by humanity through genetic engineering speeds up the process to a rate that is closer to these early instances of massive global change compared with the millennia of minute changes which happen in natural processes. Furthermore, the ability to move sections of genetic material from one highly evolved species to another is an impossibility in the course of normal evolution. It is not just a matter of doing what nature would do anyway, but creating something completely new. Quite apart from the philosophical and theological problems that this presents, the technology itself leads to shifts in lifestyle and culture, especially amongst the poor.

The links between development, environment and human justice which arise out of this technology are the subject of discussion of Chapter 5. This new situation means that a theology of creation is no longer enough to address the complex questions of environmental justice in the human community. I therefore draw into the discussion themes from liberation theology and ask how far liberation theologians have been able to address the question of justice for the earth alongside justice for humanity. So often the two are inseparable, yet an acknowledgement of the importance of environmental issues is a relatively recent development in liberation theology. I discuss the questions which arise out of environmental impact assessments and other policies which are designed to promote sustainable development. Is it really possible to put sustainable policies into practice, or are they myths or dreams?

The final chapter begins a theological reflection on the ways forward to develop a new kind of science and technology that is more in tune with human needs and the needs of the earth. I explore a theology of creation which is based on the wisdom motif, as a way of linking the transcendence and immanence of God. I examine the way wisdom has been used amongst feminist theologians and ask how far this can be applied to a renewed interpretation of an appropriate attitude to the natural world. Wisdom does not reject science as such, but draws on the philosophical tradition of Aristotle in uniting the scientific with the theological. Furthermore, wisdom theology is a theology in touch with the basic human questions surrounding environmental issues, it is essentially practical in its applications to reshaping science and technology. In our craving after knowledge in science it seems to me that we have lost sight of wisdom. Is it time to put wisdom back into the heart of science and technology? Is it time to incorporate wisdom into the metaphorical language of theological reflection?

NOTES

1. Nature philosophy is associated with German Idealism, including the work of the German poet Goethe, Georg Hegel and Friedrich Schelling. For a discussion, see J. Trusted, *Beliefs and Biology: Theories of Life and Living* (Macmillan, Basingstoke, 1996), pp. 103–13.
2. F. Capra, *The Tao of Physics* (Wildwood House, London, 1975).
3. J. Lovelock, *The Ages of Gaia* (Oxford University Press, Oxford, 1989).
4. R. Dawkins, 'A Reply to Poole', *Science and Christian Belief*, vol. 7(1) (April 1995), p. 46.
5. See, for example, A. Jensen, *Straight Talk About Mental Tests* (Methuen, London, 1981). The assumption in his work is that IQ is an accurate unbiased measure of intelligence.
6. M. Midgley, 'Strange Context: Science versus Religion' in *The Gospel and Contemporary Culture*, ed. H. Montifiore (Mowbray, London, 1992), p. 40.
7. *Ibid.*, p. 41.
8. R. Dawkins, *The Selfish Gene* (Paladin, London, 1978).
9. *Ibid.*, p. 36.
10. This view is shared by Jennifer Trusted, *op. cit.*, p. 179.
11. W. Coleman, *Biology in the Nineteenth Century* (Cambridge University Press, Cambridge, 1977), p. 112.

12. Charles Darwin's theory of evolution can be summarized as follows. Every generation of a species produces more individuals than can survive to reach maturity. There are small differences amongst individuals which leads to variety in this population. The competition between members exists because too many individuals are produced, this competition means that only those best adapted to the environment will survive to the next generation; that is 'survival of the fittest'. The characteristics which give the individuals an advantage will be passed down to the next generation and over many millennia new species will emerge. Darwin's most famous book is called the *Origin of Species*, published in 1859.
13. M. Reiss and R. Straughan, *Improving Nature?: The Science and Ethics of Genetic Engineering* (Cambridge University Press, Cambridge, 1996).
14. *Ibid*., pp. 45–7.
15. *Ibid*., pp. 60–1.
16. *Ibid*., p. 117.
17. *Ibid*., p. 164.
18. Living cells today contain enzymes which protect against the toxic effects of oxygen, which after the initial increase at the end of the Archaen era rose again in the Proterozoic era and then reached a constant ratio (that is proportion to total atmosphere) of 0.21 in the Phanerozoic era. Since this time the concentration of oxygen has remained more or less constant over many millennia. Lovelock believes that this constancy is only possible through the workings of Gaia. See Lovelock, *op. cit*., pp. 127–33.

1

Theology and biology in dialogue

What has theology to do with the environment?

A book bearing the title *Theology and Biotechnology* has first of all to give an account of itself. One response from a popular position might be: What has theology, a study of God, to do with the environment, our natural home? Such a view is based on a deep-rooted assumption that theology has little to say about the practical, material world in which we live; that for current contemporary problems the question of God is irrelevant. One purpose of this chapter is to show some of the reasons why this attitude has come into existence. Furthermore, I intend to argue that the premise on which this attitude is based is false. More important, perhaps, such an approach can act as a stumbling block for ways forward in thinking about the current environmental crisis, which threatens our existence and the survival of the planet as we know it.

It seems fair to say that we speak from a situation, rather than about it. In today's contemporary scientific and technological culture it is hard to gain the existential distance in order to look critically at what it says about ourselves. Sometimes other voices need to be listened to in order to understand what we are about. We remain blinded by the routine in which we find ourselves, oblivious to alternatives and feeling powerless to make any changes. But who are the real victims of our scientific and technological age? It seems to me that those who suffer are those who are forced to remain silent. The 'voiceless' poor and the 'silent' planet remain speechless unless we choose to listen. Their cry for justice demands that we think again about whether our cultural ethos is all that we would want it to be.

Has our narrow concentration on technological and material advancement blinded us to broader issues of human justice? Has our need for control of the world blinded us to the reality of our lack of power over natural hidden forces unleashed by our thirst for domination? It seems that emerging from this distorted ethos is unethical behaviour, both towards our fellow humans and the world in which we live. A sense of balance and interrelationship is missing.

Theology challenges us to look again at the spiritual values which serve to shape our ethos. The environmental crisis and cry for justice amongst the poor in their turn challenge theology to consider whether our perception of God has contributed to the problem in the first place. While there is a spiritual malaise felt in Western societies today, the scientific enterprise grew up in the soil of Christian beliefs and values. Some have even held the view that Christianity is to blame for the ecological crisis.[1] Even if we dispute the latter, theology can no more escape the need for a radical reorientation than other spheres. If it is considered to be irrelevant today, perhaps it has encouraged such a belief by its failure to engage with contemporary issues. The first question I intend to ask is this: what has Christian theology said about the natural world which allowed the emergence of the sciences?

Identification with or alienation from the natural world?

The paradox facing humanity is that we are both part of the natural world, yet apart from it, that somehow we understand ourselves to be different from the 'rest of nature'. Within Christian theology, also, there is a sense in which this double message comes to the surface. Paul Santmire argues that the metaphor of ascent, that is a spiritual quest for detachment, is one of escape from the natural world around us. By contrast, the metaphor of fecundity, that is the image of a land flowing with milk and honey, is an ecological motif: in other words it reinforces our identification with the land. He illustrates his point by suggesting that we could have two different attitudes in the image of climbing a mountain, depending on which metaphor is dominant:

> The movement towards the heights in one case is a movement aimed at rising above and leaving the world behind. The movement towards the heights in the other case is a movement aimed at seeing one's solidarity with the whole world, and in that it is a movement whose end is a certain poetic and affective embrace of the whole world.[2]

The ambiguity which exists is one of escape from or unity with creation. Traditional theology of the early church tended to foster one or other of these approaches. Irenaeus, for example, drew on a universal theology of creation history, rather than confining his approach simply to human sin and the grace of God. Origen, by contrast, stressed the spiritual nature of human identity and gave low status to the material world. Yet it seems to me that it is only when one or other of these strands is emphasized that problems begin. While I think that Santmire is right in identifying the ambiguity, this ambiguity is a necessary paradox if we are to avoid the danger of unwarranted identification or alienation. As I will show below and in the following chapter, the temptation for science has been towards a detachment and alienation from nature, but this alienation has come from a loss of spirituality altogether, rather than a gnostic detachment of spirit from the material. By contrast, the temptation for theology has been an alienation through a focus on history at the expense of nature or an over-identification in reaction to this felt alienation.

St Augustine is well known for his denouncement of sexuality and his seeming rejection of the body.[3] However, in his later works he seems to have come to a deeper appreciation of the value of the material world, seeing all nature as transparent to the presence of God. Augustine argued that creation is a free act of divine will, created from God but not of God.[4] For him all matter must turn to God for it to exist; if it does not it tends to nothingness. Participation in God depends on the particular calling of God to creation, coming from a trinitarian act. While his understanding of God is prescientific in the modern sense, he warned against over-identification with current science, which he believed was mere opinion.[5] While he was familiar with the scientific knowledge of the time, he believed that sufficient knowledge comes from the Creator. His model of the relation of God and creation seems a little static in a contemporary context. Even his idea of *rationes seminales*, which is the potential in things to become what they are intended to be in the mind of the Creator, suggests stability, rather than transformation. However, his stress on the *difference* between God and creation alongside the idea that all of creation is *contained* in God suggests a stable unity that avoids the danger of identification or alienation. More problematic, perhaps, are later interpretations of Augustine which drew on his earlier theology influenced by Manichaean rejection of the material world.

Aquinas is especially interesting in the present context because of the influence of Aristotle, often considered to be the father of experimental science and biology.[6] Aquinas was more aware of the limits of natural knowledge than his predecessors and sought to explain creation and its relation to the Creator from experience of the senses. The independence of created beings as such did not really find a hearing until Aquinas. However, he still held onto the Augustinian notion that all of creation would fall into nothingness without the sustaining power of God.[7] He developed the idea of secondary causes, namely that God as First Cause created secondary means through which creation existed. This allowed him to posit a freedom of creation that was difficult to ascertain in earlier theologies. For him the soul of humanity is expressed through the body, rather than in detachment from it. While this identification with nature was controversial at the time, he was still anthropocentric in his views, that is he believed that animals were made for the purpose of serving humanity. The fall of humanity led to the breakdown in human relationships with animals, so that animals now became the enemies of humans, rather than their servants. For 'things that ought to be subject to man started disobeying him as a punishment on him for his own disobedience to God'.[8]

However, he insisted that creation is good, its purpose to manifest God's goodness. Since 'creatures make up the universe in the way parts make up a whole . . . man in his entirety is for some extrinsic end, namely that he may find joy in God'.[9] While there is difference between humanity and the rest of creation in terms of hierarchical arrangement, this does not imply inferiority: 'In the same way God in the beginning made creatures different and unequal, according to his wisdom that the universe might be fully rounded. This he did without injustice and without presupposing a difference in merit in things created.'[10] It is hard to accept that his theology leads to a genuine equal merit among all creatures, since he believed that final salvation belonged to humanity alone. Furthermore, the Aristotelean idea of God as Supreme Being and First Cause, the Self-Sufficient God of the Good, put emphasis on a God who transcended the world. This existed alongside the idea of creation existing through the overflowing goodness of God, coming from Plato's *Timaeus*. Hence the ambiguity in humanity's relation with nature seems to have found impetus from Aquinas's teaching.[11] The legacy of this approach is the medieval idea of the great Chain of Being, with its double metaphors of fecundity and ascent mentioned above.

René Dubos, the microbiologist and theologian, considers that the forces which work for differentiation are more powerful than those which work for unity.[12] He insists that an untamed wilderness is alienating, our responsibility is for humanization of the world. The Benedictine approach is one which expresses a model for ecological responsibility:

> There was an ethos, we may say, of cooperative mastery over nature – neither passive contemplation, nor yet, as far as we can see, thoughtless exploitation . . . It created the existential space, we may say, in which nature could be seen in more positive terms.[13]

Dubos rejects Franciscan spirituality as one which leads to an unwarranted total identification with nature.[14] He seems to be unjustified in the harshness of his response to St Francis (1181–1225), who was above all Christocentric in his approach.[15] His love of creation stemmed from the knowledge that God loved all creation. He has achieved a greater popularity amongst environmentalists than the other saints because of his well-known affinity with animals and other creatures. Pope John Paul has described him as:

> the heavenly Patron of those who promote ecology. He offers Christians an example of genuine and deep respect for the integrity of creation. As a friend of the poor who was loved by God's creatures, St. Francis invited all of creation – animals, plants, natural forces, even Brother Sun and Sister Moon – to give honour and praise to the Lord.[16]

Yet it is easy to forget his firm adherence to a God who is Other than creation, which existed alongside his affirmation of creation. If we forget this strand then we are romanticizing his view as a total identification with the earth.

A re-enchantment of science?

Ancient science built on speculation rather than experimentation. In Aristotle's Ideal State, for example, manual labour and technology were scorned, whereas contemplation and leisure were considered the highest virtue.[17] By contrast, a Judaeo-Christian approach gave value to work as well as leisure. A scientific approach to the world was only possible in a religious context once the idea of God was separated from the world, as in Judaeo-Christian thought. Experimental science was encouraged further by the shattering of the old certainties of medieval cosmology. In other words, the discoveries of science in themselves made further scientific advancement possible. This was combined with the loss of certainty in the religious sphere as well.

Renaissance science built on the idea that all callings are in some sense 'divine'. For Francis Bacon, science is there to relieve the burdens of life which have crushed humanity ever since the fall.[18] It may be significant that a much higher proportion of the early experimental scientists of the seventeenth century were Protestants. For example, 62 per cent of members of the Royal Society in 1662 were Puritans, even though their percentage in the population was a fraction of this.[19] One of the reasons for this may have been their readiness to accept change, which was a presupposition of the newly emerging sciences. They did not see their work as contradicting their faith, rather their love for God became extended to discovering God in creation. A rationalistic philosophy seemed to encourage conceptual, theoretical approaches in science: it is through detached reason alone that we discover the workings of God. This was superseded by a voluntarist philosophy which stressed the relation of God to creation as an act of God's will.[20] This allows for some change or contingency in the natural world, thus making empiricism possible. The Book of Nature, alongside the Bible, became a means for finding God in all things.

While many of the early scientists, such as Newton, recognized that the scope of science was limited, these limitations were overlooked as successes accumulated. An almost Utopian image of the Enlightenment emerged which held that science and religion could work together for a New Age. Puritanism seemed to create the climate which favoured the growth of the freedom of science, so that it reinforced its optimism in the possibilities for the future. As Hooykaas points out: 'Puritanism and New Philosophy had this much in common: anti-authoritarianism, optimism about human possibilities, rational empiricism, the emphasis on experience'.[21] The success of scientific methods had additional effects, namely it seemed to call into question the need for theism. The original search for traces of God in all things was superseded by a desire for mastery over the natural world. The sense of human autonomy and individual freedom was a hallmark of the Enlightenment period. Gilkey puts the point succinctly thus:

> Through the disciplined use of his reason, therefore, as evidenced in the new science and its twin offering, technology and industrialisation, man saw himself in a new way as free in history:
> free to reject past authority and tradition;
> free to accumulate new knowledge of the laws of nature;
> free to control nature around him through that knowledge;
> free to transform social structures into new forms if he understood their laws;
> above all free to think what his own mind determines as the truth and his own will affirms to be right.[22]

This gave humanity a new sense of itself as creator of its own history, detached from the fatalism of Hellenistic thought or divine providence of Christian theology. However, the Enlightenment era was not entirely free from any sense of laws which exist beyond individual action. In place of the divinity of God, Nature came to be seen as 'an all encompassing hidden power, impersonal though clearly intentional, directing in a "rational" way the cosmos, man and history to their fulfillment in rational autonomy and social peace'.[23] This sums up, then, the myth of progress which is characteristic of the modern era.

Yet the naturalistic philosophy that is presupposed in this view has gradually become eroded. The instabilities of the world seem more in evidence than any sense of a power outside humanity. The old certainty that science could provide the key to human happiness and future security lingers on in some quarters, such as Frank Tipler's *Physics of Immortality*.[24] However, I am far from convinced that these dreams are any more real than the Enlightenment utopias of the Baconian era. Further, is such a future really desirable? Can a future where only the rational elements of the mind survive in any sense be called a human future?

Alongside these developments there are those who are now calling for a re-enchantment of science, that is science needs to go back to the phases when spirituality and rationality were considered together, rather than in isolation. A post-modern reaction to the modernism of the late nineteenth and early twentieth centuries can take a constructive or deconstructive approach. The latter is a simple rejection of the certainties of modernism, but nothing is put in its place. We are left in the abyss of relativism and uncertainties. A constructive post-modern approach, by contrast, seeks to revise modern views by looking back to traditional concepts in a new synthesis of old and new.

David Ray Griffin argues that we need a re-enchanted science, one that encourages a new synthesis of scientific, ethical, aesthetic and religious intuitions.[25] He argues that a simple return to a pre-modern way of life which ignores modernity threatens continuity. The scientific view, on the other hand, which assumes that disenchantment is required for scientific practice ignores the history of science: that it emerged in the context of a deep-seated faith in God. Those who are now disenchanted with science will continue to reject it unless new ways of perceiving and practising science are fostered. There is, further, a distinction between an 'enchanted' science, which can be revised, and 'sacred' science, which is immune from criticism. It seems to me that science has to re-examine itself if it is to make a positive contribution to human history. A science that is more holistic is a science in touch with its history, one that is able to accept criticism not just from within its own ranks, but from other disciplines as well. The application of science to practical problems in technology is one that seems the most relevant in the context of this book. In the past, discussions relating science and faith have tended to concentrate on physics and mathematics, to the exclusion of biology and technology.

A naturalization of theology?

The enchantment of science, where scientists were as enthusiastic about their theology as about their scientific discoveries, gave way to a secularization of scientific practice. While the desacralization of nature led to technology, the desacralization of history led to political action.[26] It is ironical, perhaps, that just as religious ideas became detached from the realms of nature and history, history began to be understood as based on natural law. The secularization of science was part of a wider detachment of religion from all human endeavour. The question which seemed most pressing was: 'How can the activity of God in social processes be understood if history as a whole and political action within it are viewed at the deepest level naturalistically?'[27] In post-Enlightenment thought, the myth of progress became associated with historical processes, so that history leads to human perfection. Darwin's evolutionary ideas seemed to introduce historical thinking into a scientific theory. Its popularity at the time may have been related to its coherence with the current optimistic attitude, which it served to reinforce. Darwin's theory of evolution seemed to do away with the need for God's providence in creation. However, the ambiguity, which was a legacy of Christian theology with respect to humanity's relationship with nature, was part of Darwin's approach as well. On the one hand, evolution seemed to be moving the world towards a progressively high ideal, as expressed in the appearance of humanity on earth. On the other hand, it seemed that humanity shared common characteristics with other animals, the primates in particular being our nearest neighbours.

What was the reaction of theologians to these developments? The success of Darwin's theories at a popular level, in spite of protestations from the church authorities, seemed to encourage two different approaches. One, taken by theologians such as Teilhard de Chardin, was to identify with the evolutionary ideas to such an extent that his theology became a grand synthesis of Christological and biological concepts. I will return to a discussion of his theology in more detail in the next chapter. The other, more common, approach was to concentrate on the dilemma of the secularization of history. As science expanded and specialized into a plethora of technically difficult and complex fields, theologians seemed to shift their attention away from a study of nature to history. David Hume effectively crushed the case for arguing for the presence of God on the basis of natural history: the traditional providence of natural theology.[28] With the demise of natural theology, an interest in nature seemed to wane as well. The Christian story, with its focus on the historical event of Jesus and based on a history of the people of Israel, seemed to be far better equipped to engage with historical issues.

The nineteenth-century liberal theologian Albert Ritschl, for example, believed that Christian faith is equivalent to God's providence in history.[29] The fulfilment of humanity is the attainment of a universal moral society based on love. In this way human potential and the kingdom of God come together through historical processes. The idea of nature was retained through understanding both history and nature as developmental processes; history became evolutionary and progressive. The onset of the two world wars contributed to the collapse of this optimism, so that:

> only sixty years after the publication of Darwin's great work this synthesis of natural science, history and theology would be rudely shattered, and the ideas of continual progress and of providence should cease to 'progress' and in fact suffer radical discontinuity and virtual extinction.[30]

The collapse in the myth of progress shattered any remaining link between history and nature, which in turn led to a detachment of theology from the scientific study of nature or philosophical interpretations of such a study. Following the demise in the optimistic liberal theology of the early twentieth centuries, the confessional theologies, such as those of Karl Barth, gained popularity again.[31] These confessional approaches did not see the relevance of issues outside the immediate individual conversion to faith in a hostile world. The gospel was to be derived and interpreted on the basis of revelation, rather than on the basis of nature or history. However, this concentration on the evangel of faith left theology isolated from the very world it sought to convert. A more recent development has been a search for an eschatological theology to answer the loss of hope in the wake of the collapse of the ideal of progress. The cry for greater relevance of faith spawned the theologies of hope of Metz, Moltmann and others.[32] These theologies introduced the political dimension into theology and were, in essence, concerned with the plight of the poor and the social oppressions of millions of people by others. Concern for individual sin and salvation was not considered to be adequate. A social conscience demanded a challenge to social structures. However, unlike their liberal predecessors, the future depended on the action of God, coming from the future rather than emerging from a historical process. In one sense these theologies represented a synthesis of Neo-Orthodox and liberal views. However, the problem which now emerges is this: has the concentration on the action of God in the future weakened a sense of realized eschatology, of the action of God in the present?

Alongside this development, in the early 1970s the growing awareness of the ecological crisis challenged theology to rethink its priorities. It was not enough for theology just to engage in political events concerned with people, the very life of the planet in the present seemed threatened by nuclear disaster and environmental collapse.[33] A purely historical theology seemed ill-equipped to deal with this crisis:

> Any theology, including our own, which takes history seriously as relevant to the promises of the gospel is going to have immense difficulties with the ecological crisis . . . Christian faith does, and must, provide a hopeful answer to this deep despair about future time latent in the ecological crisis. But surely in the light of that crisis our hope cannot be based simply on a theological answer to a historical problem, namely, that despite all the social evidence and because of the divine promise, the future will manifest a perfectly liberated society. A valid Christian hope for history cannot be a hope in a miracle, even an eschatological miracle.[34]

As a response to these developments theology was forced to become naturalized: to look again at the issues of how humans are to relate to the natural world. I am distinguishing the idea of the 'naturalization' of theology from philosophical naturalism, which I will discuss in more detail on pp. 15ff. below. I will also deal with specific ecotheologies in the following chapter. The problem with many ecotheologies is that they fail to engage radically enough with current scientific modes of thought. A further difficulty is that historical questions fade into the background: theology becomes naturalized to such an extent that its historical dimensions and its concern for humanity are gradually eclipsed. It is the premise of this book that environmental issues have to be firmly rooted not just in the context of particular environmental issues, but in the cultural, social and historical contexts as well.

Why should science listen to theology?

Given the arguments above, that there needs to be a greater rapprochement between theology and science, I could ask what real difference will it make to the actual practice of science?[35] The theoretical basis of scientific method is usually portrayed as one of detached observation and experiment. This is reinforced by the language used, the way journals are written and the multiple specializations within science which encourage an even greater distancing between the concrete objects under study and the scientist. However, in practice there is far more personal involvement than might be expected. The pioneering ideas of Polanyi, a chemist turned philosopher, show the importance of practical experience as a way of understanding the way science works.[36] Polanyi argued the case for personal knowledge, proposing that science itself involves a faith commitment. There is a difference, then, between the public image of science and the actual practice, which is the work of a human community.

A related issue is the experience of experimental science. This field includes different scientific disciplines which rely on empirical data for their investigations. In biology, at least, the division between so-called inductive science, which relies on pure observation, and deductive science, that is one which begins with a theory and then attempts to refute that theory, is not as straightforward as philosophers of science would have us believe.[37] There is also less of a distinction between science itself and application of that science in technology. Biotechnology depends on theoretical insights, but its goals and aims may be driven by an agenda that is subtly different from the detached acquisition of 'pure' knowledge.

It seems to me that no science is entirely 'value free'. The assumption that science as a social system can be separated from science as a cognitive system no longer holds water.[38] In other words social factors affect science essentially, not just superficially. Science legitimates certain social, political and economic forces. I am not suggesting that science is 'driven' from the outside, but that science cannot be divorced from its social context. The values implicit in technology are more obvious than those in 'pure' science, but values are there nonetheless. Whether these are values which the scientists themselves recognize is a moot point. Often such values are disguised by the team approach to a problem. In the solidarity of a team nobody is deemed responsible, though it is usual that one or two gain the credit in the form of publications. In this sense I could argue that the practice of the sciences is a more human effort than that of the humanities, still dominated by an individualistic approach to problems. Theologians who portray scientists as coldly detached observers have failed to notice the very human aspect of scientific research.

There are other issues which affect scientific practice which are to do with human struggles for power and dominance in a particular area. The policy in the funding of research is dominated by assessments as to whether particular individual researchers are successful; success breeds further research. The public face of science encourages us to trust science and ignore these struggles for authority. A further attitude emerges, namely trust in specialists of any subject, not just science. This has tended to leave theology to the 'experts' in their particular branch of theology. The splintering effect has made each subject alienated not just from other fields, but within its ranks as well.

The challenge of theology to science is to force science to listen to its history. From a research perspective, materials more than ten years old become the antiquated reflections of previous workers. Scientists have to be prepared to let their work become part of a body of knowledge in a way that tends to forget their individual contribution. Its culture is one which seems to give a low value to history. This is in sharp contrast to theology, which welcomes historical aspects of its faith. Indeed it is the particular history of Jesus that becomes universalized that is most characteristic of Christian theology as such. If theology has to become naturalized, then science in its turn has to become historicized; to see its achievements as part of the historical, human process. A theological interpretation of human history, which relativizes that history in the light of the future kingdom of God, can serve to challenge the practice of science as well. What are the human values implicit in its aims and goals? How far can Christian theology endorse its underlying assumptions in its treatment of nature as object?

The challenge of theology to science in a contemporary context is also to do with decisions about *which* theology in dialogue with *which* science. For the purposes of this book I will be dealing with the particular scientific issues surrounding environmental effects of biotechnology in the context of broader environmental questions, such as environmental justice and environmental ethics. The theology which seems to me most relevant to address these issues is the theology of creation from a systematic theologian's perspective. However, as many of the theological questions raised by the environmental crisis are new, the theology which emerges is a theology in the making, a theology of the environment which seeks to offer a rather different perspective from the theological deliberations of other writers.

Technology in touch with nature: philosophical guidelines

Modern science has a way of perceiving objects according to its own set of criteria. Martin Heidegger (1889–1976) described this in terms of Being, that the picture drawn by science becomes normative for the possibilities of Being; anything that is not part of this picture is non-Being. For science Being is that which is objectively defined, so that 'a Being is what is represented to man with certainty'.[39] He insists that there is another way of seeing Being, one that is 'thrust into Being' by Being itself, rather than in terms of that which is represented to humanity. In the former case humanity stands before Being and lets it be. This is different from the observation of things in themselves and their properties that was characteristic of the *experimentia* of Aristotle; or the science of the medieval period where the knowledge of things was subsumed under the dogma of faith. In the medieval period it was theology that was 'queen of the sciences'. Modern science, by contrast, projects its own conceptual framework onto beings. Mathematics is now queen of the sciences, where physics interprets the natural world in terms of spatio-temporal relations. According to this philosophy all the facts of nature can be measured by mathematics, so that more and more sophisticated calculations are all that is required. The idea of Being as *physis*, that which brings forth, in *poiesis*, that is in openness to all that is, becomes forgotten. Modern science refuses to relate to nature in openness of response. Heidegger insisted that science needs to be shed of its pretensions to dominate nature, rather than rejected as such.

Since Heidegger's time new developments in physics, especially cosmology, have led to a new sense of awe and mystery, a resurgence of the mystical nature of the universe. F. Capra's *The Tao of Physics* is an important example of this work.[40] But I wonder what kind of mystical experience is suggested by this new physics? In one sense it shows the limits of science's attempt to control the universe. The *physis* of Being has reasserted itself again. While the new physics may inspire the religious consciousness of humanity, it is not recognizable as Christian faith. It seems closer to a generalized spirituality of place, that is not critical of human actions, but emerges, as it were, from within the natural world.

However, when it comes to applied sciences, especially technology, this development of a new physics seems to me to be largely forgotten. Technology draws on the Newtonian models of science which were familiar to Heidegger. He rejected the idea that technology is neutral, but according to the modern concept is a means to a specific human end.

Aristotle identified four 'causes', that is the 'formal' cause or design; the 'material' cause or matter itself; the 'efficient' cause, that is in the mind of the craftsperson and the 'final' cause or purpose. Cause for the Greeks was all of these and related to the *techne* of human action. For Heidegger there is a fundamental meaning of cause that is related to the Being itself or Ultimate cause. In modern understanding of technology, cause means only the instrumental use to humans, that is, the efficient cause overshadows the others and the only cause given worth is that which produces an effect. Instead of the original purpose of *techne*, which means 'bringing forth', modern technology has led to a challenging, or *herausfordern*. This 'puts to nature the unreasonable demand that it supply energy that can be extracted and stored as such'.[41]

Modern technology in this way 'sets upon' nature in a way that makes it go against its original intention. There is a denial of that which is manifest in Being itself. This is not simply an application of science, but a forcing of nature to deliver what humanity requires. While in science there is a pretence of the ontological independence of the objects of science, in technology even this vanishes.[42] While science describes nature as theorized objects, in technology nature becomes a resource to used for the benefit of human goals and aims and part of the will for power and domination.

Heidegger believed that the essence of technology which demands that nature is a 'standing reserve' of energy, or *Enframing*, arose in the period *prior* to the modern physics of the seventeenth century and the modern technology of the eighteenth century. In other words, technology preceded modern science and seems to be shaping Western history. The illusion that everything is a construct of humanity begins to take its grip. All other possible kinds of revealing from within the natural world are effectively blocked out by this approach. It delivers, further, an ambiguous message. On one hand, there is a frenziedness in the ordering process, which blocks every other view and so endangers the relationship to the essence of the truth. On the other hand, it seems to lead to a structured ordering that allows humanity to exist in apparent security and so give the pretensions of truth.[43]

There are two possible attitudes we can take once we realize that technology has taken over all aspects of life. One is to try and master technology, to 'put it in its place'. The other is a total rejection of technology as evil. Heidegger opts for the first alternative. The hope for technology is not so much in giving up technology altogether, but becoming aware of its foundations and presuppositions. In this way, 'Within and beyond the looming presence of modern technology there dawns the possibility of a fuller relationship between Man and Being – and hence between Man and all that is – than there has ever been'.[44]

An important question concerning the above is how does Heidegger's view relate to philosophical naturalism? In traditional understanding, this is the philosophical approach, popular half a century ago, which asserted that the world can best be accounted for by the categories of natural science, including psychology and biology, without recourse to the transcendent. Heidegger seemed to be most concerned to address positivism, which elevated physico-chemical explanations. However, it seems to me that he also moved beyond naturalism in that he rejected the idea that the purpose in science and technology was self-evident from the sciences themselves. He also offers a more sophisticated approach than supernaturalism, which is based on the idea of divine intervention. Dewey believes that a supernaturalist approach, which is antinaturalist, led to some of the 'bloodiest events in history'.[45] Philosophical naturalism is not inevitably atheistic, but it rejects the idea of God until it can be proved by empirical evidence.[46]

Has there been a 'failure of nerve' in Western civilization along with a 'loss in confidence in scientific method'?[47] While this may be true in the popular mind, technology seems to carry on in a relentless fashion which shows anything but a loss of confidence. Yet Hook is right to detect a loss in confidence in science, even while there is an inevitability to the technological process which seems to heighten human anxiety about the future. Hook's suggestion that all we need is ever more application of scientific method puts him firmly in the naturalist camp. Naturalism is, nonetheless, more sophisticated than simple materialism. Whereas materialism aims towards material goals, naturalism aims at a discovery of our place in the natural world. Yet it seems to me that naturalism on its own does not offer an adequate critique to technology as such. While it may be fair to say that science alone cannot be blamed for social ills, to assume that current scientific practice is blameless seems to me to be naive. A reformulation of science in the manner implied by Heidegger's philosophy offers a radical, but realistic approach. Radical because it attacks the fundamental attitudes embedded in science and technology. Realistic because it does not reject technology outright, but sees a real promise and way forward in spite of our current context.

Those who would consider themselves to be naturalist in contemporary culture are also those who stress humanity's unity with nature, rather than those who adopt the scientific method. This brings some ambiguity to the label 'naturalist'. According to the former view, human continuity with the world is stressed so that nature becomes an all-inclusive category, while still asserting that religious ideas are the products of human imagination. If we regard those who use ecology as a basis for their philosophy as 'naturalists', such as in so-called 'deep ecology', science and 'reductionism' of any kind are rejected. Identification with the natural world seems the logical outcome. The ambiguous relationship between humanity and the natural world seems to assert itself once more, with a tendency towards alienation or identification. Those who are naturalist from within a traditional scientific perspective, such as Hook, seem to retain an essential alienation from nature. Those who are naturalist from within an ecological perspective seem to move towards identification with nature. In one sense ecology has been called a 'subversive' science, since its focus on interconnectedness goes against the stream of traditional 'reductionist' approaches of the scientific method. The power of Heidegger's approach is that a thin line is drawn between alienation and identification, since by looking behind to the purpose of Being itself, detachment is still required. Yet it is a detachment that is empathetic, rather than hostile. It is furthermore compatible with theological approaches. For example, Paul Tillich draws on Heidegger's philosophy and identifies Being with God, who is also the God of history and nature.[48]

Technology in touch with nature: theological metaphors

The language which we use to describe God is necessarily oblique: God is described through symbols, either through analogy of like with like or through metaphor, that is spotting a thread of similarity in an otherwise dissimilar image. The language of science uses the language of metaphor in order to help us grasp its concepts. In molecular biology words such as 'recognition', 'avoidance', 'mistakes' pepper the scientific literature. However, this introduces a subtle, covert idea of purpose into the biological process. The language leads to subtle shifts in our reliance on the explanatory power of molecular biology. By speaking in words which are familiar to us through the use of anthropocentric terms, it seems to speak as if there is an intention in biological processes. The non-intentionally that is part of the scientific method becomes subsumed under this intentional language. One good example of this is the so-called 'selfish' gene.[49]

It is entirely understandable that biologists will seek to put their discoveries into a language that makes sense to us. The difficulty is that this language is then used by sociologists to apply biology to human behaviour. Social biology does just that, it draws out the 'lesson' for human social action from biological processes.[50] But the 'folk psychology' of the biological work gives a 'spurious aura of real explanatory content'.[51] It seems to me that the application of biological processes as a model for behaviour makes a category mistake and is circular in argumentation; what is forgotten is that the language used is metaphorical rather than literal. If we base our hope on the seeming progression in biology, we forget that on its own biology has no such teleology or purpose, rather it has been introduced by the scientists in order to help us understand a concept. This is one of the reasons why naturalism on its own is doomed to failure, it can never really escape an essentially anthropocentric approach.

Theology, on the other hand, is more at home in the world of symbol, analogy and metaphor. According to the apophatic tradition God is known as much by what we do not say, as by what we do say. In other words we can never spell out in a literal way who God is and what are the inner purposes of the Deity in specific detail. All our descriptions must remain provisional, while recognizing that belief in the goodness of God is a fundamental starting point and act of faith. Traditional ways of thinking about the transcendence of God put God 'above' the world, in 'contrast' with the world. This tends to split the action of God from the action of the world; God becomes a primary being in a cosmological hierarchy, a spatial God above and the earth beneath. Plotinus's characterization of creation as emanation from God removes the idea of transcendence as contrast with the world. The problem now is that the distinction between God and the world becomes blurred. Furthermore, the model is essentially static, rather than dynamic.

Modern theologians sensitive to the current climate of hopelessness and lack of purpose, following the collapse in the myth of progress, have sought to express the transcendence of God in terms which emphasize God as a God of the Future. God is a God who transcends creation in a temporal, rather than a spatial sense. The metaphor of God as sovereign above the world, exerting power and authority, gives way to a metaphor of God as enabler and persuader of creation. The providence of God becomes displaced to temporal, rather than spatial categories. A biological metaphor of perpetual cycles of death and new life is challenged by the linear model of progression of time characteristic of historical processes. However, this linearity is itself challenged by God coming from ahead, a return *from* the future. A theological approach can draw on the biological metaphor of life after death while still challenging its circle of return. Furthermore, a theological approach can draw on the model of historical processes, while coming from ahead as well as offering continuity with the past.

The theological metaphor which is fundamental to Christian experience is the idea of God as personal and moral being. This image seems, in some ways, to distance us from the natural world, since the context of human life becomes political; human relationships and ethical action become fundamental. Yet the idea of God as personal and of humanity defined in relationship to God safeguards our sense of human responsibility. Kaufman suggests that the idea of God as personal seems to encourage a view of creation as material for human purposes, even if the idea of responsibility for creation in the motif of stewardship counters this tendency. In this way: 'the very ideas of God and humanity as they have gradually been worked out over the millennia, are so framed as to blur or even conceal our embeddedness in the natural order as we are increasingly conceiving it'.[52]

Yet I would challenge the idea that viewing God as personal necessarily leads to a depreciation of the natural world, even if historically the religious significance of the natural world was weakened subsequently. If anything the idea of a personal God can encourage human involvement with creation since God is seen from the perspective of the incarnation: God is in human solidarity with us and with creation.

St Francis of Assisi took this notion to heart in his poetic appreciation of human solidarity with God and with all creatures. Nature itself can give us metaphors for human living, the moon and sun are our 'brother' and 'sister'. The metaphor for the world is familial, personal relationships include those with all creatures. St Francis regarded all of the natural world as sanctified by the incarnation. He has been described as a 'nature mystic' in the sense that all of nature was the specific context which led to his ecstatic experience of God. He is the patron saint of all those who: 'taking pleasure, sometimes ecstatic delight – in the exuberance and diversity of Creation, thankfully regard them as expressions of divine splendour, sacramental intimations of glories beyond human apprehension'.[53]

St Francis's love of nature was intensified by his detachment from worldliness and material acquisitiveness. The intensity of his appreciation of all of nature as a metaphor for the love of God is hard to imagine without an equal appreciation of his ascetic lifestyle and his commitment to the poor and disadvantaged. The Celtic saints showed a similar combination of the love of nature with commitment to service. The early biographers of St Francis had a poetic and hagiographic licence to exaggerate stories and draw on the mythical accounts derived from the Irish saints.[54] The mixture of legend and fact is a characteristic of Celtic Christianity.[55] The literal truth or otherwise of these stories is not really an important issue. The most important point is their function as metaphorical stories, namely to foster love and pity for all of creation and for the Creator.

Furthermore, the particular cultural context of the time needs to be taken into account in order to assess the significance of these stories for the ecological crisis today. For example, in medieval times souls of the dead and angels were thought to reside in birds. Even in earliest Christianity the song of birds was thought to be worship.[56] The stories of St Francis's encounter with birds, such as his duet with the nightingale, his preaching to the birds and larks singing near his deathbed makes sense in this context. For St Francis it was self-evident that animals had rights; their dignity came from value in the sight of God. There seems to be little evidence for Francis's empathy with fish or reptiles. However, he did not allow the appearance of reptiles to undermine his compassion for them as God's creatures, any more than natural human repulsion towards lepers.[57] The story of the taming of the wolf at Gubbio most likely portrays consummation of the dream of reconciliation with nature. However, his fearlessness of wolves in general is probably authentic. The wolf represents forces of brutality, greed, lust and power-seeking. Its animal instincts are overcome by virtue. The histiography or otherwise of the story misses the point of the hagiographers.

St Francis reminds us of the unity of interests of God, humanity and nature. The stories help us to begin to empathize with the interests of creation, while keeping human needs as a priority. Reptiles remind us of our duty towards lepers, not the other way round. It seems to me that St Francis is important not just in the sense of introducing a deeper affinity with all of the natural world, but he is important also because he challenges us to see ourselves in the context of the cosmos. A simple return to 'nature' worship is not enough as it fails to address human responsibility. As I mentioned in an earlier section, those who believe that St Francis supports such an orientation have misinterpreted his views.

Have we lost the metaphor of the cosmos as a way of thinking about our place in the world? The meaning of cosmology has shifted away from questions about the place of humanity in nature, to overall questions about the structure of the physical world. Cosmology as an all-inclusive term seems to be out of fashion. Stephen Toulmin suggests that we need to re-examine our theology in the light of traditional cosmology. He suggests that we need:

> to swim against the tide, to suggest that we should look again at the case for reinstating cosmology, in its older and broader sense as a field of discussion that overlaps the boundaries of science, philosophy and

religion; and for regarding physical cosmology as a sub-branch, not just of physics, but of general cosmology in this older and broader sense.[58]

However, a return to cosmology does not mean a return to the science of the Babylonian era. The science of cosmology can exist side by side with cosmology as a field of view. This bears some parallels with the idea of ecology, which is both a science and an ideology. He comments:

> It is no accident, for instance, that the term ecology embodies the same kind of ambiguity as cosmology itself. On the one hand we have ecology the science, involving the study of the food chains and other natural systems that link together the fauna and flora of the small ponds, and so maintain the harmonious equilibrium on which their continued lives depend. On the other hand there is ecology the movement, involving the defense of natural habitats and systems against disruption as a result of ill-considered industrial and social policies and campaigns of public education and political action to promote that defense.[59]

Ecology, as a metaphor for theology, is distinctive in its natural affinity with cosmology and a global perspective. Biological science can lead to two different metaphors for philosophy and theology. 'White' philosophy is rooted in psychotherapy and teaches a doctrine of self-command and detachment. Human motivation and self-fulfilment are core elements in this approach. 'Green' philosophy, on the other hand, is rooted in ecological science and lays emphasis on the unity of the human species with the rest of the world, ethically we are to live in harmony with nature. While the 'white' approach bears some affinity with the Epicureans, the 'green' approach bears some resemblance to the Stoics. These 'white' and 'green' approaches in their extreme forms lead to alienation from or identification with the natural world, which I discussed earlier. It seems to me that St Francis is both white and green, white in his detachment from the material world and green in his sense of unity with the cosmos. Nonetheless, overall his theology seems to encourage a rather passive view of nature: we look at nature and through observation celebrate its communion with God in union with ourselves.

St Benedict, on the other hand, introduces the idea of active participation with creation. He is also important in his insistence on the vow of stability. In contrast with the itinerant lifestyles of the Franciscans, the Benedictine Rule established the notion of permanent communities. It seems to me that the Benedictine model of stability is even more significant than his notion of wise stewardship of the land. Stability implies identification with place and with the land, which has its roots in the Jewish faith. In the biblical tradition the fertility of the land depends on the obedience of Israel. Misuse of the land is easier if that place is thought of in terms of a temporary dwelling which can be discarded once all the 'resources' are utilized. Stewardship combined with stability allows for the establishment of relationship, not just with the individual creatures, but with the place itself.

Israel's identification with place is emphasized by the traumatic experience of exile. The wilderness in Israel's tradition is ambiguous: it is place both of God's action and human rebellion. Cohn suggests that ambiguity is characteristic of periods of transition, the ambiguity of the movement from slavery to land ownership in a 'landless' zone.[60] Mount Sinai acts as a symbol of the unity between heaven and earth. It represents the beginning of time, when Israel was created through the giving of the law in a way that is analogous to bringing order out of chaos. Cohn believes that 'Israel always retained its self-conscious memory of being a created, covenanted, not a "natural" people. It remembered its creation at Mount Sinai.'[61] The God of Israel is a God of the covenant, but it is covenant in the context of the land.

Traditional theological approaches have tended to see the land as the backdrop against which the covenant takes place.[62] In other words the metaphor for God's action is covenant, then creation. The ecological crisis has challenged theologians to rethink their assumptions in subsuming creation to covenant in this way. Creation is part of the covenant of God to all living creatures, as symbolized in the Noahic account.[63] Eventually Zion replaces Sinai as the place of Yahweh's abode. In Christian theology the covenant with Christ replaces the covenant at Sinai, but the image of Zion as the symbol for the final community remains, at least in the background. Jerusalem has remained the 'mother' city to which Jewish people have looked with nostalgia, but in the exile the Torah took over its function as a focal centre. The sense of sacred place became ambiguous through the exile experience and in this way relativized the importance of Jerusalem. All major religions have the exile as a metaphor for religious experience. This in itself, tends to foster a sense of 'cosmic homelessness', which draws believers away from their perception of this world as their home.[64]

For Christians Christ replaces the function of city and temple as the ultimate place of God's dwelling. Those Christian theologies which stress Christ and Christ's future coming as the only basis for faith could act to sever the human link with the land. Yet the idea of pilgrimage remains and continues through the centuries of Christianity. The challenge for theology today is to foster a sense of the value of place without regressing to a 'primitive' animism. Some theologians have suggested that the most appropriate term to use in a post-modern culture is not God, but the Sacred.[65] This, to my mind, fosters theological metaphors of place: in other words it associates divine action with space, rather than time. The problem with the metaphor of the Sacred is the tendency towards a narcissistic focus on the natural world in a way which forgets human distinctiveness. Feminist theology has tended to stress human symbiosis with nature, to counter supposed 'masculine' rationality of technology. Lasch argues that narcissism is a reaction to technology as such, rather than a specifically male and female approach. In other words:

> The technological project of achieving independence from nature embodies the solipsistic side of narcissism, just as the desire for mystical union with nature embodies its symbiotic and self-obliterating side. Since both spring from the same source – the need to deny the fact of dependence – it can only cause confusion to call the dream of technological omnipotence a masculine obsession, while extolling the hope of a more loving relation with nature as a characteristically feminine preoccupation. Both originate in the differentiated equilibrium of the pre-natal state, and both, moreover reject psychological maturation in favor of regression, the 'feminine' longing for symbiosis no less than the solipsistic 'masculine' drive for absolute mastery.[66]

While Lasch has constructed a case against some feminist theologies, he is unnecessarily scathing in his sweeping criticism that this is a 'handful of shopworn slogans and platitudes'.[67] He advocates respect for nature, but I have my doubts as to whether his style encourages a respectful attitude. I will be returning to the issues raised by feminist theology in more detail in Chapter 2.

Santmire believes that one way forward is through a process of *ingression*, that is, staying within the framework of technological society and waiting for signs of a new beginning.[68] The difficulty here is that there seems to be an assumption that these signs will come from within technology itself. There is little sense of a radical reappraisal of the technocratic structures as suggested by philosophical approaches of Martin Heidegger referred to above. It seems to me that the challenge for theology is to introduce an eschatological dimension into the idea of the sacred so that it is in touch with changes in science and technology. The cosmos can become a realm in which God's future action appears.

The motivation for concern for the natural environment then becomes both sacramental and future-orientated. Yet the seeds of this future are in a promise yet to be fulfilled. Haught suggests that 'When looked at eschatologically its value consists not so much of its sacramentally mediating a divine "presence" as of its nurturing a promise of future perfection'.[69] A purely sacramental theology finds difficulty in including destructive forces at work in creation. An eschatological approach puts perfection in the future and so is not alarmed by inconsistencies in the present. Yet a hampering of that future leads to despair, we are turning away from the divine promise. The advantage of Haught's interpretation of nature as promise is that it suggests a way in which we can look to the future in a positive way. A possible disadvantage is whether the notion of 'progress' becomes reintroduced by way of 'process' thought. How can we distinguish between the future generated by human manipulation and that which is part of the divine promise? Is there a sufficient understanding of the need to 'let things be' which emerges from Heidegger's analysis? In other words, I am aware that while a reinstatement of cosmological and eschatological approaches in one sense serves to foster concern for the ecological crisis, there is a temptation for complacency in our critique of the current situation. The need for a *metanoia*, a turning away from either a passive narcissistic unity with nature or an aggressive control of nature, is lost. It seems to me that a theology of the environment has to include a strand which is ready to admit fault and sin, death and decay, both in the human and non-human realm.

In the following pages I will be drawing out in more detail some of the issues raised in this chapter. In particular, in the next chapter I will be seeking to address how far different theologies of creation can give us guidelines in a construction of a theology of the environment. In Chapter 3, I will look at the different ways our treatment of nature has been discussed according to environmental ethics. In particular, how does theology challenge philosophical approaches to environmental ethics? Is there an approach which is theologically valid and philosophically sound? In Chapter 4 I will look in more detail at the questions and issues raised by the growth in biotechnology. It seems to me that there has been an over-concentration on environmental issues such as climate control or pollution. While these have immediate effects on the environment, the potential effects of biotechnology are staggering. Another advantage of the discussion of biotechnology is that it leads naturally into issues of justice in the human community, which I will address in Chapter 5. How far can theologies which emerge from situations of oppression instruct new approaches to the environment? What is the relationship between liberation theology and environmentalism? In the final chapter I will draw the threads together from the previous chapters and give an outline of a theology of the environment which takes into account the recent developments in biotechnology, ethics and theology.

NOTES

1. Lynn White is the most notorious advocate of this view, which I will return to again in the following chapter. See L. White, 'The Historic Roots of Our Ecological Crisis' *Science*, vol. 145 (1967), pp. 1203–7.
2. H. P. Santmire, *The Travail of Nature: The Ambiguous Ecological Promise of Christian Theology* (Fortress Press, Philadelphia, 1985), p. 25.
3. See, for example, J. Moltmann, *The Way of Jesus Christ* (SCM Press, London, 1990), p. 85.
4. C. J. O'Toole, *The Philosophy of Creation in the Writings of St. Augustine* (Catholic University of America Press, Washington, 1944), p. 13.
5. *Ibid.*, p. 63.
6. E. Mayr, *The Growth of Biological Thought* (Harvard University Press, Boston, 1982).
7. T. Aquinas; C. O'Brien (ed.), *Divine Government: 1a–2ae 103–109*, vol. 14 (Blackfriars, Cambridge, 1975), question 104, p. 39.
8. T. Aquinas; E. Hill (ed.), *Man Made to God's Image: 1a 90–102*, vol. 13 (Blackfriars, Cambridge, 1963), question 96, p. 123.
9. T. Aquinas; W. A. Wallace (ed.), *Cosmogony: 1a 65–74*, vol. 10 (Blackfriars, Cambridge, 1967), question 65, p. 12.
10. *Ibid.*, p. 13.
11. H. P. Santmire, *op. cit.*, pp. 85–95.
12. R. Dubos, *A God Within* (Charles Scribner's Sons, New York, 1972), pp. 27–8.
13. H. P. Santmire, *op. cit.*, p. 79.
14. R. Dubos, *op. cit.*, p. 167.
15. E. A. Armstrong, *Saint Francis; Nature Mystic: The Derivation and Significance of the Nature Stories in the Franciscan Legend* (University of California Press, Berkeley, 1973), p. 219.
16. Pope John Paul, 'Peace with God the Creator; Peace with All of Creation', *Theology in Green*, issue 3 (July 1992), pp. 26–7.
17. R. Hooykaas, *Religion and the Rise of Modern Science* (Scottish Academic Press, Edinburgh, 1972), pp. 76–7.
18. *Ibid.*, pp. 92–4. I will come back to the ideas of Bacon and the changing attitudes to nature in Chapter 2.
19. *Ibid.*, p. 98.
20. E. B. Davis, Jr., *Creation, Contingency and Early Modern Science: The Impact of Voluntarist Theology on Seventeenth Century Natural Philosophy* (Indiana University, PhD thesis, 1984), pp. 1–13.
21. R. Hooykaas, *op. cit.*, p. 143.
22. L. Gilkey, *Reaping the Whirlwind: A Christian Interpretation of History* (The Seabury Press, New York, 1976), p. 194.
23. *Ibid.*, p. 196.
24. F. Tipler, *The Physics of Immortality* (Macmillan Education, London, 1995).
25. D. Griffin, 'Introduction: Postmodern Proposals', in *The Re-enchantment of Science*, D. Griffin (ed.) (State University of New York Press, New York, 1988), pp. 1–46.
26. L. Gilkey, *op. cit.*, p. 198.
27. *Ibid.*, p. 199.
28. R. W. Hepburn, 'Hume, David' in *A Dictionary of Christian Theology*, A. Richardson (ed.) (SCM Press, London, 1969), pp. 162–3.
29. A. Ritschl, *The Christian Doctrine of Justification and Reconciliation* (T. and T. Clark, Edinburgh, 1900); see also L. Gilkey, *op. cit.*, pp. 213–15.
30. L. Gilkey, *op. cit.*, p. 216.
31. See Chapter 2 for a more detailed discussion of Karl Barth's theology and its significance.

32. For example, J. Moltmann, *Theology of Hope* (SCM Press, London, 1967); R. D. Johns, *Man in the World: The Theology of Johannes Baptist Metz* (American Academy of Religion Dissertation Series 16, Scholars Press, Missoula, Montana, 1976).
33. Doomsday books predicting total ecological collapse were particularly common in the earlier writings of environmentalists. See, for example, J. Schell, *The Fate of the Earth* (Chaucer Press, Picador, London, 1982); for discussion, see M. Palmer, *Dancing to Armageddon* (Harper Collins, London, 1992).
34. L. Gilkey, *op. cit.*, p. 237.
35. For additional discussion, see C. Deane-Drummond, 'Biology and Theology in Conversation: Reflections on Ecological Theology', *New Blackfriars*, vol. 74, no. 865 (October 1993), pp. 465–73.
36. M. Polanyi, *Personal Knowledge* (Routledge and Kegan Paul, London, 1958).
37. K. Popper, *The Logic of Scientific Discovery* (Hutchinson, London, 1958).
38. D. R. Griffin, 'Introduction' in D. R. Griffin (ed.), *op. cit.*, pp. 1–46.
39. H. Alderman, 'Heidegger's Critique of Science and Technology', in *Heidegger and Modern Philosophy: Critical Essays*, M. Murray (ed.) (Yale University Press, Newhaven and London, 1978), p. 36.
40. F. Capra, *The Tao of Physics* (Wildwood House, London, 1975); F. Capra; *The Turning Point: Science, Society and the Rising Culture* (Wildwood House, London, 1982).
41. M. Heidegger, 'The Question Concerning Technology', in *The Question Concerning Technology and Other Essays*, trans. and with an introduction by W. Lovitt (Harper Torchbooks, New York, 1969), p. 14.
42. H. Alderman, *op. cit.*, p. 46.
43. M. Heidegger, *op. cit.*, pp. 33–5.
44. W. Lovitt, 'Introduction', in M. Heidegger, *op. cit.*, p. xxxvii.
45. J. Dewey, 'Naturalism in Extremis', in *Naturalism and the Human Spirit*, Y. H. Kikorian (ed.) (Columbia University Press, New York, 1944), p. 6.
46. S. P. Lamprecht, 'Naturalism and Religion', in Y. H. Kikorian (ed.), *op. cit.*, p. 36.
47. S. Hook, 'Naturalism and Democracy', in Y. H. Kikorian (ed.), *op. cit.*, p. 40.
48. For a discussion of the relationship between God and Being, see P. Tillich, *The Courage to Be* (Collins, Glasgow, 1952).
49. R. Dawkins, *The Selfish Gene*, 2nd edn. (Oxford University Press, Oxford, 1989). For a theological critique of Dawkins, see M. Poole, 'A Critique of Aspects of the Philosophy and Theology of Richard Dawkins', *Science and Christian Belief*, vol. 6, no. 1 (1994), pp. 41–59.
50. Most notorious of social biologists is E. O. Wilson. For a review of his work, see C. Barlow, 'From Ants to Anthropology: E. O. Wilson', in *From Gaia to Selfish Genes*, C. Barlow (ed.) (Massachusetts Institute of Technology Press, Cambridge, 1991), pp. 147–61.
51. A. Rosenberg, *The Structure of Biological Science* (Cambridge University Press, Cambridge, 1985), p. 262.
52. S. D. Kaufman, *The Theological Imagination* (Westminster Press, Philadelphia, 1981), p. 226.
53. E. A. Armstrong, *op. cit.*, p. 17.
54. *Ibid.*, p. 44.
55. See, for example, J. P. Mackey, 'Introduction: Is There a Celtic Christianity?', in *Introduction to Celtic Christianity*, J. Mackey (ed.) (T. and T. Clark, Edinburgh, 1989), pp. 1–25. For further discussion see C. Deane-Drummond, 'Recalling the Dream: Celtic Christianity and Ecological Consciousness', *Theology in Green*, issue 7 (July 1993), pp. 32–8.
56. E. A. Armstrong, *op. cit.*, p. 65–7.
57. *Ibid.*, pp. 171–2.
58. S. Toulmin, 'Cosmology as Science and as Religion', in *On Nature*, L. S. Rouner (ed.) (University of Notre Dame Press, Indiana, 1984), p. 28. See also S. Toulmin, *The Return to Cosmology* (University of California Press, Berkeley, 1982).
59. *Ibid.*, 1984, p. 36.
60. R. L. Cohn, *The Shape of Sacred Space* (American Academy of Religion, Scholars Press, Chico, California, 1981), pp. 14–20.
61. *Ibid.*, p. 57.
62. P. Schoonenberg, *Covenant and Creation* (Sheed and Ward, London, 1968).
63. R. Murray, *The Cosmic Covenant* (Sheed and Ward, London, 1992).

64. J. Haught, *The Promise of Nature* (Paulist Press, New York, 1993), p. 40.
65. H. P. Santmire, 'Epilogue: the Birthing of Post-Modern Religion', in *Critical Issues in Modern Religion*, R. A. Johnson, E. Wallwark with C. Green, H. P. Santmire and H. Y. Vanderpool (eds) (Prentice Hall, Englewood Cliffs, 1973), p. 454.
66. C. Lasch, *The Minimal Self* (Picador, London, 1985), p. 246.
67. *Ibid.*, p. 248.
68. H. P. Santmire, *op. cit.*, 1973, pp. 458–50.
69. J. Haught, *op. cit.*, p. 110.

2

Theologies of creation: past and present

I argued in the last chapter that history has dominated contemporary theological inquiry to the extent that any interaction with modern scientific issues has become marginalized. The 'naturalization' of theology stemmed from a reaction to this historicism in a way that continued to ignore scientific discourse. I suggested that theology can become relevant to the creation of new ways of perceiving science and technology. Moreover, such changes in theology need to become part of a shift towards a more holisitc understanding of the place of humanity in the cosmos.

In this chapter I intend to examine different theologies of creation in the light of changing attitudes to the natural environment. Before the mid-twentieth century, under the influence of Darwinism, the environment tended to be viewed in a biologically deterministic way as the implacable Power of Nature.[1] A culture that is aware of environmental issues or an environmental culture gradually emerged in the 1970s beyond seeing the natural world as something 'out there' to be managed for human benefit. Environmentalism as a cultural phenomenon seems to have shifted once again in the 1980s and 1990s from this 'grass-roots' movement to an institutionalized, potentially alienating, language culture.[2] In this case the idea of the environment includes a political dimension and views humanity primarily in dialectical economic relationship with it. In this respect, it has returned to a more objective 'modernist' approach, undermining the more inclusive language of popular culture. The term 'nature', like the environment, is also an ambiguous term since humans are part of nature, as well as distinct from nature. Historically the term 'nature' has changed in meaning and as it has about fifteen current meanings it is a highly ambiguous term.[3]

I will be using the term 'natural environment' to refer to non-human nature. The main questions I wish to address in this chapter are: How have attitudes to the natural environment changed in the early modern period? Does this exploration give us any clues as to the roots of the environmental crisis? How are contemporary theologies of creation of relevance to these issues?

In 1967 the historian Lynn White pointed to Christianity as the root cause of the environmental crisis. He suggested that the command in Genesis to have dominion over the earth was a licence to abuse the planet, so that 'we shall continue to have a worsening ecological crisis until we reject the Christian axiom that nature has no reason for existence except to serve man'.[4] Most scholars recognize that his seemingly exclusive blame of Christianity is oversimplified. His view can be refuted both on the basis of a faulty exegesis of Genesis, as well as the simple observation that Christians are not solely responsible for causing environmental damage.[5] However, there is still a lingering doubt that in some way Christianity may have contributed to a negative attitude to the natural world by its anthropocentric or, as feminists would claim, androcentric position.[6] Recently, the philosopher Stephen Clark has suggested that 'though Christians cannot be exclusively to blame, and plenty of non-Christian countries have done as badly or still worse, it may still be true that "the Christian axiom" prevents a resolution'.[7] He acknowledges that seeking out and blaming certain groups, or alternatively romanticizing about golden eras where people lived in harmony with nature, is naive and ignorant of the complexity of history. The successive changes in history may be linked more to ordinary human need for survival than any particular theological doctrine. However, it is still possible to ask if Christian theology has in any way blinded us to the consequences of environmental damage. If our doctrine is that the world is for human benefit alone, or that the material world is one which we must abandon in favour of the spiritual realm, then there will be little reason for self-restraint.

Changing attitudes to the natural environment in the early modern and modern period

While theologians have pointed to the sacramental quality of the natural world as a Christian affirmation of the environment, there is still a suspicion that some theologies have in some way contributed to the current crisis. I will be examining the Reformers Martin Luther and John Calvin, as even if Christianity *per se* is not blamed, in more popular writing the finger is still pointed at Protestant theology and Calvinism in particular.

Martin Luther (1483–1546) was inclined to view the natural creation as a 'concatenation of hostile energies' under a curse following the fall of humanity. The wrath of God expresses itself in the evil which we see in the universe, which has a way of becoming 'a kind of existential springboard for grace'.[8] The radical separation of nature and grace in Luther's doctrine of the two kingdoms tended to depreciate the material world. This compares with Aquinas's more synthetic treatment of nature and grace.

Other aspects of Luther's theology were more affirming towards creation. He insists, for example, in the original goodness of creation before the fall of humankind.[9] He allows for the present value of creation as that which displays a 'mask of God'. In this sense all creation has a miraculous quality, as it acts like a window to its Creator. While he affirms all creation, the highest praise is still reserved for humanity. He believes that the earth was created by God as a suitable dwelling place for humanity 'into which he is brought by God and commanded to enjoy all the riches of so splendid a home'.[10] In particular, the intellectual quality of humans is reserved for humanity alone since 'a gleam of the life of the intellect does not exist in other earthly creatures . . . the human being in his mind, soars high above the earth'.[11] In the ideal state before the fall, humankind exercised dominion over all creatures, but in such a way that there was no need for cunning or skill. After the fall the dominion exercised became inferior and after the deluge there was further deterioration in the relationship between humans and creation. He comments, somewhat curiously: 'I believe our bodies would have been far more durable if the practice of eating all sorts of foods – particularly, however, the consumption of meat – had not been introduced after the deluge'.[12]

Luther values creation as part of God's good workmanship which will be encompassed in the future redemption. The positive value of creation becomes clear in Luther's commentary on Psalm 8 where he states that 'on that day everything will become new and beautiful'.[13] It seems that both heaven and earth are somehow caught up in the new existence since humans have, seemingly, a choice of home: 'we shall be wherever we like – in heaven or on earth . . . a broad and beautiful heaven and a joyful earth'.[14]

Luther is, overall, affirming of the non-human world, yet he is unashamedly anthropocentric in his views. This is reinforced by the theological focus of justification by faith which tended to push consideration of the created realm of nature to the periphery of his thought. However, it would be unfair to suggest that his views encouraged an active hostility towards creation, or that it was the privilege of humans to exploit the earth entirely for their benefit.

John Calvin (1509–64) believed that all knowledge of God came from scripture. His portrayal of God is as one who is sovereign over both inner and outer historical events. The rule of God over external forces finds expression in his doctrine of divine providence, while God's rule over inner life finds expression in the doctrine of election.[15] This emphasis on salvation history leads to a more dynamic view of history and helps to overcome the more static view of God as one who upholds the principle of order. Our freedom lies in our willing acceptance of what is ordained by God such that all that we achieve is by an experience of *sola gratia*.[16] Calvin's focus is on the power of God, who is the absolute Creator of all that exists and who continues to exercise mastery over all creation. He distinguishes fate, which is blind, from divine providence, which expresses the purposes of God and through which human beings are fulfilled. There is a general providence which operates over all of the created order and a particular providence which operates in individual lives. For Calvin human freedom is the willing acceptance of God's will. There is an ambiguity in Calvin's thought which seems to make human beings responsible for evil, but not responsible for good. In addition, while on the one hand his stress on individual salvation tended to foster autonomy and individualism, on the other hand his ideal of humans becoming instruments of God's glory encouraged a Christian socialism where the requirements of the individual were subordinated to the needs of the community and the requirement for social justice.[17]

A similar ambiguity exists in Calvin's approach to creation. While he affirms the high place of humanity as part of the divine purpose, there are also strands in his theology which show a more sacramental view towards the natural world. The natural creation becomes the theatre which displays the glory of God and shows the awesome beauty of the Creator.[18] Nonetheless, Calvin encourages humanity to reach out for the transformation of the world which tends to weaken any more contemplative strands in his thought. In a similar fashion to Martin Luther, Calvin's overriding concern for soteriological issues tends to push his comments on nature and creation into the background of his thought.

Even while Calvin was still alive in 1554, Gomez Pereira put forward the view that animals were not so much creatures of God, but machines or automata. Humans differed in that they had a mind or soul. René Descartes (1596–1650) popularized this view, which was widely accepted, possibly because it gave a Christian rationalization for the harsh treatment of animals. There were a few dissenters to this position; Henry More, for example, described it as a 'murderous doctrine'.[19] The visible mechanization of society gave a visible analogy for these writers. One of its more unfortunate effects was that it served to justify harsh treatment of those people who were on the margins of society who were treated little better than animals.[20]

There was a shift in the underlying philosophy of the natural world from the Greek view, dominated by Aristotelian ideas, to what Collingwood has termed the 'Renaissance' view.[21] According to Greek philosophy the world of nature is saturated by mind, so that it has both a 'mind' and a 'soul' in which individual organisms participate. The Greek view was based on an analogy of being between the human person and the natural world. The mechanistic view was influenced by the work of Copernicus (1473–1543), Telesius (1508–88) and Bruno (1548–1600). They denied that the natural world was an organism and insisted that it was devoid of both intelligence and life. Its movements were imposed from the outside by the 'laws of nature'. As Collingwood points out:

> Instead of being an organism the natural world is a machine: a machine in the literal and proper sense of the word, an arrangement of bodily parts designed and put together and set going for a definite purpose by an intelligent mind outside itself.[22]

Francis Bacon (1561–1626) championed an anthropocentric vision where he perceived that all of the natural environment was at the disposal of humans.[23] He insisted that scholars left their detached speculations and applied their knowledge to relieve the burdens of life. He believed that the mechanistic view of the world of nature freed Christianity from the pagan belief which deified nature and identified God with creation. Humankind's mastery over the world in the arts and sciences can repair some of the evil consequences of the fall. In a cultural atmosphere that was buoyant from new discoveries and international travel, Utopia seemed to be within our grasp. This optimistic vision tended to cloud any sense of the corruption of humanity. There was little sense of the threat posed by the upsurge in population or the power of potential forces for change.[24]

Keith Thomas suggests that the historical period between 1500 and 1800 was crucial in setting the stage for both the development of an intense interest in the natural world alongside doubts about our relationship with it. We are left with a legacy of current anxiety.[25] There was a growing development in modern historical consciousness alongside this ambiguous relationship with the natural world. The more static views of both nature and history characteristic of Greek philosophy were eroded by a number of different factors including:

(a) medieval concepts of apocalyptic where forms could be changed by divine *fiat*;
(b) Renaissance humanism which criticized inherited structures;
(c) reformed theology's challenge to medieval ecclesia;
(d) Calvin's notion of divine providence working through history;
(e) scientific method, which assumed contingency in the natural world.

The rise in experimental science, according to historian Langdon Gilkey, was the most important factor in breaking the Greek concept of changeless forms.[26] However, the two factors are likely to have had a mutual, synergistic effect on each other. In other words, the gradual shift in the perception of the natural world fostered the development of empirical science. A growing historical consciousness was part of the cultural matrix which allowed the growth of Darwinian biology nearly a century later.[27]

Immanuel Kant (1724–1804) thoroughly endorsed the mechanical view of the natural world championed by Descartes and Isaac Newton (1642–1727). He understood the physical world as consisting of immutable, hard and dead conglomerations of moving particles.[28] For Kant 'nature' is the proper object of scientific study and appears to humans as regular and predictable, though the 'thing in itself' is not the same as the mechanism. This marks a shift from the strictly mechanical view where the mechanism was thought to be reality. Kant rejected any idea that God was the proper object of scientific study; rather God is a necessary postulate which serves practical reason and an idea which helps humanity to understand the unity of the natural world. Humanity differs from the rest of the natural world through our freedom, which overcomes the deterministic quality characteristic of non-human existence.[29]

Kant's philosophy has been described as an 'ecological sieve' to Lutheran and Calvinistic theologies.[30] Their peripheral treatment of creation issues was subsequently lost in the post-Kantian era. Alongside this influence there was a radical change in the perception of the natural world in the nineteenth century. The concept of the natural world characteristic of the modern era is based on an analogy between the changing flux of human affairs studied by historians and the processes of the natural world.[31] The emerging historical consciousness was based on the central paradigms of process, development and change, and encouraged the belief that new forms could emerge.

Hegel (1770–1831) marks the transition period to the modern era. He incorporated the idea of becoming or *process* into the primary form of logical becoming. This is not a movement in time or space, or a change of mind, but a logical movement that is part of concepts as such.[32] Unlike Plato, the system of concepts is subject to growth instead of being static. The logical becoming in the world of concepts leads to changes in the natural world. The significance of humanity is that it becomes the vehicle of the Mind, which is the precursor to the being and becoming of the Spirit.[33] For Hegel both the natural world and mind are presupposed by the Idea, which like the natural world has objective reality. This contrasts with Kant who believed that our mind creates the natural world. For him the postulate of God is required simply for practical reason. Kant's philosophy is a subjective idealism, while Hegel's is an objective idealism.

The theories of Copernicus offered an explanation of the world which seemed to do away with any need for God in a fixed medieval cosmology. Natural theology, which finds evidence for the existence of God in the natural world, then retreated at first to biology as giving evidence of the wise Designer. William Paley (1743–1805), for example, and other natural scientists gloried in biological science as a way of disclosing the wisdom of God. His influence was weakened by the pioneer of modern biology, Charles Darwin (1809–82). Darwin's publication of the *Origin of Species* challenged natural theology by explaining the intricacies of nature in terms of natural causes.[34] The two theological issues most as stake were:

(a) Darwin seemed to treat humankind as simply one of the animals;
(b) his views seemed to remove any need for God in the world.

Darwin's theories reinforced a more vital concept of the natural world that helped to break down the dualism of mind and matter characteristic of the mechanical Cartesian view of Descartes. Collingwood comments that 'it is fair to say that the concept of vital process, as distinct from mechanical or chemical change, is here to stay and has revolutionised our conception of nature'.[35]

Darwin's hypothesis left a double legacy in its combination of historical consciousness with biological change. Positively, it seemed to encourage a very different, more vital concept of the natural world that culminated in the philosophy of Bergson and aligned the modern view with the ancient Greek concept of the world as an organism. Negatively, the distinctions between God, humanity and nature start to become blurred so that pantheism seems the most obvious option for those wishing to retain a theistic view.

Post-Darwinian responses: Karl Barth (1886–1968) and Teilhard de Chardin (1881–1955)

Karl Barth's dialectical theology

Karl Barth swam against the cultural stream of his day in rejecting all attempts to root theology in history, believing that this amounted to 'culture Protestantism' which carried the force of 'man' rather than God. He rejected his teacher's support of the 1914 war policy which seemed to him to underwrite the bankruptcy of their theology for genuine Christian ethics.[36] What is the significance of creation in Karl Barth's theology?

The starting point for Barth's doctrine of creation is an affirmation of belief in God the Creator, rather than an apologetic position against the purely materialistic understanding of the world.[37] He presupposes that God exists in relation to creation which is both distinct from God, yet exists as a result of God's action and will. These assertions find their basis in faith in Christ, which is impossible to demonstrate or contest. The Creator and creation exist through the action of God's free grace.[38] Heaven and earth are gifts coming from God's free creative acts and stand in creaturely relationship with God.[39]

For Barth the doctrine of creation emerges from the self-witness of scripture understood Christologically. Jesus Christ, as God and human, teaches us that God, though absolute, has a partner outside 'himself'. Jesus Christ, as human and God, teaches us that humanity is never absolute, but is in creaturely relationship with God. The significance of Christ for creation goes beyond this noetic sense in that there is an ontological relationship between Jesus Christ as the Word, through which God made, upholds and rules creation.[40]

Creation is bound up with salvation history in Barth's theology. The creative work of God is envisaged as a trinitarian action, with the work of the Father, the Son as the Word and the Spirit as one who gives life.[41] God's election in Christ and through him the whole of humanity precedes any discussion of the creation of the world out of nothing. Barth describes the story of Genesis as a saga, which is a deliberate stress on the intuitive and poetic account of pre-historical reality. Barth describes the animals as 'inferior' to humanity, their value comes both from prefiguring humans and as sacrifices prefiguring the sacrifice of Christ.[42] The dominion of human beings over the animals is not an unrestricted lordship, but tempered by the supreme lordship of Christ and an affirmation of the goodness of creation.

Barth's theology of creation is thoroughly anthropological. However, he remains positive in his assessment of creation, even if it is because creation is bound up with the covenant between God and humanity. The incarnation of Christ stresses the high value that God gives to creation. Both covenant and creation together affirm the 'Yes' of God the Creator.[43] For Barth, the reality of creation stems from an awareness of its creaturely dependence on God. There is nothing to add to creation, it is perfect as created by God:

> The only thing which can be better than creaturely existence is the goal of the covenant for which the creature is determined in and with its creation. But in the order of created existence as such there can be nothing better than what is . . . Even its future glorification presupposes that it is already perfectly justified by the mere fact of its creation.[44]

It is well known that Barth strongly rejected the natural theology of Emil Brunner.[45] However, he does allow for a 'luminosity of creation' as that which shows itself to be the creation of God. The self-revelation of God does not depend on the brightness of the created world; rather all the joys and sorrows of creation pale into insignificance compared with the exaltation and humiliation of Christ.[46] Nonetheless, the lights of creation are not extinguished by Christ's light or the sinfulness of humanity. He believes that these lights of creation are 'their own revelations' and come from the creature itself.[47] Human recognition of these lights depends less on faith as on simple 'common sense'.[48] The revelation of God in the incarnation is not 'alien' to the creaturely world, but is an affirmation of the one declaration of the Creator. While the dominant movement in Barth's theology is from creation to covenant, the converse is also true, that is Christ is the seal of the sustaining and protecting work of God in creation.

Teilhard de Chardin's evolutionary synthesis

Teilhard de Chardin, like Karl Barth, reacted against his teachers. However, his was a reaction against a conservative form of Roman Catholicism that seemed to take little account of modern scientific theories. His theology is in stark contrast with Karl Barth in that he stresses the immanence of God in creation, rather than God's transcendence.

His theology was heavily influenced by Charles Darwin's hypothesis. However, he extends this theory so that the axis of evolution is towards *hominization* and ultimately *Christologization*. He introduces a number of terms which have since failed to gain credibility in biological science. The biological term *orthogenesis* meant direction in evolution, regardless of external influence. Teilhard uses this word to describe the totality of successive stages of evolution which tend towards increasing complexification.[49] This process begins with *cosmogenesis*, or the physical organization of matter, followed by *biogenesis*, or the emergence of life, then the development of a nervous system or *cephalization* and finally the emergence of self-consciousness or *hominization*. Once we extend the idea to human society as a whole, we have now reached a period of *intensification* where we become aware of ourselves as evolved beings. The final goal of orthogenesis is Omega, which Teilhard believes we can deduce from science. It merges with Christian theology in its goal of the cosmic Christ.

In contrast with Karl Barth, Teilhard uses evolution to provide him with a natural theology. Both writers tend towards an anthropocentrism, though the basis for this in Barth's theology is his strong sense of the covenant, while for Teilhard it stems from his perception of evolution as culminating in the emergence of humanity. He modifies the biological theory by suggesting that the driving force in evolution is not chance and necessity, but 'love energy' which today can serve to unite the human community. His aim was to unite the modern humanistic idea of material progress, which he describes as 'headless', with a Christianity that had become too detached from the world and was like a head without a body.[50] He believes that science needs a soul if it is to become personal rather than impersonal. The creation, incarnation and redemption are part of one movement where the incarnation serves as a link between Christ the evolver and Christ the redeemer. In this sense redemption is more like a *re*-creation than individual forgiveness, found in Barth's thought.

Teilhard's vision points to an acceptance of the material world through a pervading sense of the love energy of God within the whole cosmos in a kind of panpsychism. He moved beyond the theology of his day in giving a new model of God as one who is within creation and ahead of creation. This future is conditioned by human action in history, which places a far greater weight on our responsibility within the world, as well as a greater optimism of our potential to achieve this goal. While his theology corrects the more individual focus in Barth's thought, it tends to be over-optimistic, especially in subsequent interpretation. Teilhard's vision was particularly vulnerable to attack in the wake of the collapse of liberal humanistic optimism. Nonetheless, he has had a wide influence, especially for process theology, which I will discuss in more detail below.

Contemporary Roman Catholic responses: Karl Rahner (1904–64) and Hans Küng (b.1928)

Karl Rahner was heavily influenced by the transcendental philosophy of Joseph Maréchal. He uses his philosophy to reconcile the Christian belief in God as Creator with modern culture. He broadens the notion of grace to include our general capacity to transcend our environment. The potential for grace is built into our beings as a transcendental peculiarity. It is so much part of us that even if we reject God's grace, it still becomes determinative for our existence.[51]

Rahner moves away from any dualism of nature and grace which implies that grace is 'added' to our natural beings. Scholastic theology proposed the idea of grace added to human nature. Rahner rejects this view as unwarranted 'extrinsicism' based on an archaic philosophy orientated towards mechanical physics.[52] For Rahner, grace operates in our creaturely being and prepares us for the supernatural end reached, in part, through justification and ultimately through the beatific vision.[53] Yet it is God's intention to give the whole creation a supernatural end. The line between natural knowledge based on creation and the revelation of God becomes blurred. The distinction between nature and grace is for methodological clarity, as it is a reminder that grace is entirely gift rather than based on merit. However, nature and grace still remain 'distinguishable entities'.[54] Creation and redemption are no longer distinguishable, so that redemption includes creation in the sense that the whole is related to the part. The miracle of divine love is expressed most fully when God becomes incarnate in Jesus Christ.[55] This demonstrates that, unlike Karl Barth's theology, the priority for Rahner's Christology is the incarnation, rather than the cross.

The purpose of creation is still for the sake of humanity, which he understands as the 'highest being in all creation'.[56] The natural order has somehow a 'crying need for grace' and as such is always more than simply natural being.[57] The cross reminds us that the future of creation is not Utopia on earth, while the decision of God to enter this world in the incarnation leads to the belief that the future of the world is already decided in terms of salvation. This promise of future salvation acts like a mandate for humanity to work responsibly for a unity between creation and redemption. Hence, while Rahner retains the clear distinction between God and creation, his notion of a single movement of nature and grace overcomes the dualism implicit in scholastic theology. As a consequence all of creation is affirmed and saved.

Rahner's all-encompassing concept of the relation between God and the world fosters a dialogue with all branches of human life, including that of science. He is well aware of the problem of the specialization and fragmentation of science. It leads to a sense that the specialization is such that it is quite impossible to achieve a comprehensive view of reality, which he terms a 'gnoseological concupiscence'.[58] Theology differs in that it puts its focus on transcendence, including human transcendence. Science needs theology as a reminder of the subjectivity of science and as a means to reflect on a unity of reality which is greater than the sum of the parts. The failure of much science is its tendency to monopolize other sciences and its failure to listen to other approaches, including theology. The task of theology becomes one of persuading science to consider its human nature, rather than offering any material contribution. This modest goal is more realistic than Teilhard's grand synthesis, though I am unsure if Rahner's transcendental aspect of human nature is really necessary for his reflection on theology's critique of science. In other words, the specifically theological contribution remains rather unclear.

Hans Küng, like Rahner, aims to include the secular world as a serious part of his theological discourse. In fact he believes that a theology which refuses to face scientific questions is an expression of either arrogance or ignorance.[59] He fears that traditional Roman Catholic faith has isolated itself from the developments in science and has become trapped in the medieval scholastic view of the world. While science in a methodological sense has to leave out God, the individual scientist needs to retain an 'open-mindedness about reality as a whole'.[60] Like Rahner, he believes that the problem of finding a coherent unity is connected with the question of God, as such this means that we need a 'new, a modern understanding of God'.[61]

Küng prefers a panentheistic notion that God is in the world and the world is in God: God is the infinite in the finite and the absolute in the relative. He insists that the distance between God and creation is irremovable, both in the ontic sense of divine being and the noetic sense of divine knowledge. He prefers the notion of a transpersonal God, rather than God as non-personal or a supreme person and this reflects his panentheism. Like Karl Barth, Küng believes that a positive attitude towards the world stems from belief in God as Creator. The world once again becomes meaningful, as does the place of humanity in it: 'Believing in the Creator God of the world thus means accepting with greater seriousness, greater realism and greater hope my responsibility for my fellow men and for the environment and the tasks assigned to me'.[62] He is well aware of the revolutionary significance of molecular biology within biological science and addresses the question of chance and necessity in the work of Jacques Monod (1910–76) and Manfred Eigen (b.1927). For Monod the concept of chance seems to predominate to such an extent that the world becomes meaningless, while for Eigen the world is an 'inescapable necessity' coming from biological laws. Küng resists the temptation to put God in the 'gaps' of human knowledge. Rather, he looks to the philosophical issues implied by modern biology. He believes that we face a choice between humanity in isolation where there is no God or an affirmation of the unity of all reality in God.[63] Küng redefines God as one who is both infinite, yet intimately bound up with the suffering of the world. Both humanity and the environment are bound up together into the being of God, so that God is not:

> without feeling, incapable of suffering, apathetic in regard to the vast suffering of the world and man, but a sympathetic, compassionate God, who, in the future, changing everything by liberation from sin, suffering and death leads to infinite justice, to unbroken peace and to eternal life – a God of final redemption.[64]

Küng has tilted the relationship between God and creation into the future, so that the power and love of God is realized in the future. The idea of God as one who contains the world and shares in its suffering marks a radical departure from the traditional hierarchical schemes which use monarchical imagery.

The contribution of Eastern Orthodoxy

The Orthodox perspective is one which sees 'nature' as encompassing the whole of reality, including humans. The otherness of God is stressed so that God remains 'indefinable, ineffable and incomprehensible'.[65] Orthodoxy draws heavily on the writings of the early church fathers, especially Gregory of Nyssa's theology of God and creation. God and creation are related through the notions of *diastema* and *metousia*. *Diastema* is the unique dimension of space and time in creation, which acts between the being or *ousia* of creation and the ontologically separate Creator. There is no *diastema* directed between the Creator and the creation since the whole of the creation is permanently co-present with the Creator. The distinction between creation and Creator is set by the limits of space and time. *Metousia* acts in a more positive way and refers to the participation of creation in the *energia* of God. There is no possibility of desacralization since the creation would cease to exist if it failed to participate in

God. *Diastema* is the basis for the transcendence of God, while *metousia* expresses God's immanence.

Humanity has a key place as mediator between God and creation. Christ has broken the *diastema* between creation and God, so that he shares the *ousia* of God. Through Christ, humanity can move across the limitations of space and time into the eternal presence of God. Human freedom is preserved in that humans are given the choice as to whether they participate in this goodness.

In the Russian Orthodox Church of the nineteenth century, a theological position known as Sophiology developed, drawing on the writings of Vladimir Solovyov (b.1853), Sergius Bulgakov (b.1871) and Pavel Florensky (b.1882). The writing of Bulgakov was subsequently attacked by traditional Orthodox theologians, such as Vladimir Lossky.[66] His basic premise was that Sophia, or wisdom, is the intelligible basis of the world, that is the wisdom of nature.[67] He was radical in his suggestion that Sophia is part of the *ousia* or being of God. Sophia is like the link between God and the world, as divine Sophia expressing God's transcendence and as creaturely Sophia expressing God's immanence.

More recent contemporary Orthodox theologians have restated the importance of Christian theology in confronting the ecological crisis. John Zizioulas insists that the ancient teaching of the church that creation is *ex nihilo*, that is has a beginning, is still relevant today.[68] A beginning implies that the world also has an end, that 'taken in itself it hangs in a void, and cannot avoid the threat of death'. The world, in and of itself, does not possess any way out of the limitations of space and time and the subsequent nothingness implied by this fact. However, the Christian faith affirms a God who created the world in love, so there must be some means for the survival of the world, an escape from this threat to nothingness. The problem for theology is how to speak in such a way that it remains coherent in normal scientific and philosophical discourse. The way forward is to seek a way of uniting the world and God, that is finding a link between the two without abolishing the natural otherness between God and creation.

Zizioulas believes that the solution to the problem is through the place of humanity in creation. For him: 'It is in the human being that we must seek the link between God and the world, and it is precisely this that makes Man responsible, in a sense the only being responsible for the fate of creation.'[69] He rejects the idea, widespread since the Middle Ages, that human beings are superior to the rest of creation because of our rational capacity.[70] He believes that the danger of a rational interpretation is that the world can be rationalized as having its existence simply for human benefit. This is at the heart of a selfish attitude to the creation and is one of the causes of the ecological crisis. The contribution of Darwinian biology was that it showed that rationality is not exclusive to the human being, but other creatures also possess a measure of consciousness and reason. According to current philosophy, human beings are unique in their ability to create a world of their own with culture, history and so on, not just as a means of survival, but as a way of expressing their identity. This implies that the most distinctive human characteristic is not so much rationality, but freedom. Yet human freedom is still limited by the possibilities available at the time. This differs from the absolute freedom of God who creates out of nothing and is not bound by a given reality. However, as humans are made in the image of God we still strive towards the absolute freedom. As such the human being is confronted by tragedy which 'is the impasse created by a freedom driving towards its fulfilment and being unable to reach it'.[71]

The unique capacity for human freedom is linked with the sinfulness of humanity. The classic interpretation of the fall was that human beings exceeded the limits of the freedom given to them by God. However, according to St Irenaeus the fall of Adam consisted of the use of his absolute freedom in the wrong way. This implies that repentance is a redirection of absolute freedom, rather than a restriction on the use of freedom. Zizioulas now asks: How can creation transcend its limits and take part in this drive to absolute freedom? He believes that the human capacity to reach out for absolute freedom is the means of hope of survival for all creation.

The means through which this task is achieved is through humans: 'By taking the world into his hands and creatively integrating it and referring it to God, Man liberates creation from its limitations and lets it truly be.'[72] The model of the proper relation of humanity to the world is through the person of Christ, who acts as supreme Priest of Creation. As humanity offers the bread and wine to God in the Eucharist, so do we recognize the source of all creation comes as a gift from God. Through this act, creation comes into relationship with God and is transformed from its natural limitations into a bearer of life. Creation acquires a sacredness as it no longer spells death, but life exercised through the free choice of human action. It builds an ethos, a way of being, which counters the cultural crisis of which the ecological crisis is one aspect.

Vincent Rossi articulates an explicit ecologically orientated Orthodox theology, so that for him 'to be a Christian is to be an ecologist'.[73] Like Zizioulas he believes that the ecological crisis is part of a cultural problem, he specifically points to modernity as that which caused 'an unprecedented psychological and spiritual rupture of the people from its cultural and spiritual roots'.[74] He links the kingdom of God with the re-establishment of humanity's role as priests of creation. This combines a serving attitude with deep respect for creation. He believes that this is a more fruitful approach than the popular idea of humans acting as stewards of creation, which usually results in simple management and is ultimately utilitarian.

Radical perspectives: process theology and feminism

While Hans Küng and Karl Rahner move beyond traditional Catholic formulations of God, the theologians of process take the secular view of the world with even greater seriousness. Process theology is bolder in its willingness to reformulate both its image of God and the relationship between God and the world. It stems from the empirical method in theology and philosophy that became popular in North America in the 1920s. Alfred North Whitehead (1861–1947) combined empiricism with the organicism of Bergson into a comprehensive metaphysical system.[75] There are two main general characteristics of his philosophy. Firstly, Whitehead believed that all of reality has a subjective element, thereby weakening the distinction between mind and matter. The only difference between subject and object is a temporal one, so that objects are past events, while subjects are present events. Hence subjectivity becomes part of matter itself. Secondly, he rejected the assumption in the biological understanding of perception that cause and effect are simultaneous. Instead, our observations have a 'physical pole', or objective reality, prior to our reception of them as empirical data. The decision that we make to give a sensory experience a quality, for example the 'greenness' of grass, introduces a novel element into the experience known as the 'mental pole'.[76]

Whitehead's philosophy has a number of theological consequences. Human beings become part of inanimate matter in such a way that subjectivity is felt throughout the universe. The idea of the eternal in God's consequent nature is emergent from the universe. Eternal objects become 'possibilities' for realization. Charles Hartshorne and John Cobb used Whitehead's philosophy and redefined it in Christian categories.[77] The 'otherness' of God is through a modified version of panentheism. Process theology seeks to develop a view which emerges from the concrete rootedness in the world. It identifies more readily with ecological issues compared with traditional theology. Cobb believes that the ecological issue is more fundamental than that of social justice, since it affects the survival of the human race. He argues in favour of an ecological attitude, which he contrasts with the technological attitude that seems to set no limits to human use of natural resources. He draws on the theology of Albert Schweitzer as one who encouraged the love of all living things, but was active in the practice of medicine. He believes that St Francis of Assisi was too far removed from our civilization and too passive to be of any real use to theology today. Instead, a theology of the natural world 'need not prevent man from continuing his scientific investigation of nature . . . but it would prevent ruthless indifference to the consequences of our actions for the living environment'.[78]

He believes that the idea that subjectivity is at the heart of the universe is vindicated by science itself. He claims that since we have 'excellent evidence that electromagnetic events do take account of their environments, it is reasonable to affirm that they are instances of unconscious feeling'.[79] I am less convinced by this aspect of his argument. Even though matter, in particular its electromagnetic properties, can share the sensitivity to the environment of mental processes, this does not lead to the conclusion that they are instances of 'feeling'. While it seems reasonable that in the past the dualistic attitude to mind and matter was too sharp, some distinctions remain.

For Cobb, our basic commitments arise or emerge, rather than happen through choice. God is the God who gives life and love through the Spirit. It is belief in this Spirit that acknowledges that the human person is not alone that is the basis for hope in the future. While a solidarity with the emergent Spirit is a source of comfort in our difficulty, it is hard to imagine how such an emergent Spirit could have any real influence in shaping the future of the world and that of the cosmos. Moreover, process theology seems to retain a deep optimism about the future of the planet and an ideal of progress where the notion of God's Spirit has replaced the Omega of Teilhard de Chardin. As such, it is less challenging to the scientific technological world view than Cobb seems to suggest through his notion of ecological thinking.[80]

Like process theology, feminist theologians seek to reformulate our understanding of God by beginning with the experience of the world. The particular experience is that of patriarchy and sexual oppression of women. The relationship between God and creation is reformulated so that the idea of God as the transcendent one who has power over the universe is replaced by God who is immanent in the universe. Rosemary Radford Ruether, for example, begins with the notion of the Holy Spirit, who is 'the ground of being of creation and the new creation'.[81] She believes that the distinction between mind and matter has its roots in patriarchy, which attempted to put man and mind on a supernatural level and woman and matter on a natural level. Eventually apocalyptic literature emerged which envisaged a radical denial of the world, so that the 'patriarchal self-deception about the origin of consciousness ends logically in the destruction of the earth'.[82] She argues that a denial of the harmony in nature goes hand in hand with the repression of women. A radical reshaping of the underlying values of society would include a much more active role for women which goes beyond the merely symbolic presence that they have had in the past. She insists that the false polarities of male and female need to be overcome before there is any hope for a change in attitude towards the natural environment. While liberation from sexist oppression is a natural counterpart to liberation from a tyrannical attitude to the natural world, Ruether's argument that feminist theology is required as a precursor to a revised attitude to the environment seems rather weak.

For Ruether Gaia is the new symbol for the feminine divine.[83] In *God and Gaia: An Ecofeminist Theology of Earth Healing* she takes in an eclectic mixture of sources from biblical studies, patristic theology, feminism, cosmology, biology and social anthropology. Gaia is defined as the 'living and sacred earth', while God is the 'monotheistic deity of the biblical tradition'.[84] She includes a discussion of biological evolution, seemingly as a creation story in parallel with other theological accounts. She is harsh in her treatment of science, believing that mechanistic thinking forces us to ignore the complex interactions in the natural world. The healing of the dualism between mind and matter, nature and grace, comes through revisiting themes in the Christian tradition such as covenant and sacrament. She identifies God with the covenant theme and Gaia with the sacrament theme. However, the two sources of the divine are intimately connected. At the heart of the Godhead the idea of interrelationships predominates: 'Thus what we have traditionally called "God", the "mind" or rational pattern holding all things together, and what we have called "matter", the ground of physical objects, come together . . . coincide'.[85]

Other feminist theologians, such as Dorothee Soelle, Shirley Cloyes and Ursula King, include an explicit political manifesto in their theology.[86] For Soelle and Cloyes the starting point is a creation in need of liberation, rather than creation as a gift of God, which is characteristic of more traditional theology. While they are influenced by process theology, they view the divine and human interrelationship as dynamic and reciprocal, rather than emergent as in process thought. They aim to replace a technological attitude, which they believe leads to despair, with a 'passionate love for creation'.[87] Once human labour is separated from creation it leads to an alienation from the earth and a retreat into religiosity. We become deaf to the 'cry of the earth'. Ursula King argues that a way forward is to develop a spirituality which allows us to experience a sense of harmony between ourselves and the world around us. She draws on the Eastern religious traditions which stress the bond between ourselves and the earth.[88]

Elizabeth Moltmann-Wendel believes that our reformulation of God in feminine categories is a prerequisite to an adequate theology of creation. She draws on the Jewish belief in the feminine presence in Shekinah for inspiration where 'The *Shekinah* is something like the cosmic, reconciling, earthly side of God which accompanies Israel into exile'.[89] The feminine images of God have persisted as a kind of 'subculture', even though they have faded in the written tradition. The symbolism of milk and honey is an example of this latent feminine strand in Jewish writing. As with many other feminist theologians, Moltmann-Wendel's starting point is the experience of interrelationships. This includes a sense that sociality is at the heart of all things. The communal aspect gives room for a basic love of self that is unashamed of emotional experiences. This can lead to a rediscovery of friendship with the natural world which evokes a sense of anger and pain as the creation is not yet whole. For: 'If I am whole, I suffer the fact that creation is not whole, and have the right and duty to protest that creation is no longer whole.'[90] This view leads to ecological activism.

Christopher Lasch as cultural analyst has criticized feminist theology's identification of femininity with the earth and masculinity with domination of nature.[91] He believes that both are narcissistic and that a politics of conservation must rest on a firmer philosophical foundation than a mystical adoration of nature. While there is an element of truth in his criticism of the 'shopworn' slogans that have filtered into the women's movement and the environmental and peace movements, the idea that feminist theology is automatically associated with mystical adoration of nature and loss of individuality seems exaggerated. Sallie McFague, for example, reconstructs our image of God so that God is mother and the world is God's body.[92] She resists the charge of pantheism, claiming that her theology is panentheistic, for 'while we, as members of the body, are radically dependent upon the life-giving breath from the spirit, God, as the spirit, is not so dependent upon the universe'.[93]

Post-modern synthesis: Jürgen Moltmann

An ecumenical theology of creation which has had increasing influence has come through the discussions and writing of the World Council of Churches (WCC). The idea of the 'integrity of creation', first coined at the Vancouver conference in 1983, was joined with concerns for human justice and the socio-political implications of the ecological crisis. The overall negative assessment of the effects of technology was echoed in a number of different theological approaches, ranging from process theology to Eastern Orthodoxy.[94] The proposed solutions to the problem are different. For Birch this is the 'personal' encounter with the universe, where the distinctions between God, humanity and the natural world become blurred. For the Malawian Harvey Sindima the starting point is the interrelatedness of the human community and its rootedness in the land.[95] The African concept of creation and the cultural belief in the power of life, 'moya', remains intact as the 'organising logic of the African world'. He believes that the radical disruption of the rhythm of life through the myth of progress is the root cause of the ecological crisis. Feminist theologians are critical of the unfair bias given to the views of the Orthodox Church at WCC meetings.[96]

While there is some consensus in terms of a critical appraisal of theological issues, the diverse theological starting points lead to a rather bland common denominator approach. Ronald Preston has noted the difficulty of linking an elusive idea, such as the 'integrity of creation' with the prime aim of the WCC towards social ethics, liberation theology and sexist oppression.[97] He suggests that theologians have lost their nerve in their outright rejection of technology, arguing that greater weight should be given to those who argue for a constructive use of technology.

Lutheran theologian Jürgen Moltmann has written extensively on the ecological crisis and has formulated his doctrine of creation in ecological categories.[98] It seems to me that his approach is post-modern in that he seeks to move beyond the myth of progress characteristic of the modern era, while adopting a highly eclectic view which draws on the insights of both contemporary and classical theology. He believes that the anxiety which pervades our attitude to the environment stems from our failure to arrive at an adequate understanding of God.

He believes that our first task is to image God as one who loves creation, rather than think of God in categories of power. God is the God of the social Trinity, whose existence is through the fellowship of Father, Son and Spirit in dynamic interrelationship.[99] He links the doctrine of *perichoresis* of the Eastern Orthodox Church with the idea of interconnectedness at the heart of reality that is characteristic of feminist theology. Drawing on process theology, he envisages a reciprocal, but not equivalent, relationship between God and creation. He departs from process thought both in his belief in creation *ex nihilo* and his belief in the future of creation coming from the God of the future, rather than emergent from creation. He also believes that the theme of self-emptying or *kenosis* is at the heart of the life of the Trinity.[100] This Christological term is broadened in his theology so that it becomes the initial act within God prior to the creation of the world. This has the further effect of bringing the idea of suffering into the heart of the Godhead. Moreover, Moltmann is keen to stress that God is not impassable, but is one who shares our suffering and this suffering reaches right into the heart of the Trinity.

Moltmann envisages all of the natural and historical worlds participating in the life and glory of the Godhead as a parallel theme alongside that of the social Trinity sharing in the suffering of creation. Like Karl Barth, Moltmann rejects the notion of 'epiphany' faith, yet he allows for natural theology in his eschatology. The vision of God in earlier schemes has its place in the future hope. He claims that '*theologia naturalis* is at bottom *theologia viatorum* and *theologia viatorum* will always concern itself with the future *theologia gloriae* in the form of fragmentary sketches'.[101]

The image of the sabbath is one which extends to the future and only in the future is the glory of God manifest through the shared participatory life of all creation. In the human community our role as *imago Dei* precedes our future vocation to become *gloria Dei* through *imago Christi*. The inner life of the social Trinity becomes the model for the human community, which is marked by love and fellowship with the oppressed. The oppressed includes those marginalized in society as well as the earth itself. The natural world is beloved of God, so it follows that humanity, as God's image, is obligated to love creation. Human beings 'love all their fellow creatures with the Creator's love. If they do not, they are not the image of the Creator, and the lover of the living. They are his caricature'.[102] The relationship between God and humanity is a special one, humanity both acts as a mirror of God and is God's viceroy on earth.

The twin ideas of the cosmic Christ and the cosmic Spirit stress the immanence of God in creation. God as cosmic Christ reaches out to the groaning and suffering of creation in a way that gathers up the natural world into Christ's death and resurrection. Christ's resurrection marks the beginning of the new creation and the end of death itself.[103] The cosmic Spirit pervades every aspect of earthly creation in a way which refuses to split the material from the spiritual. After the resurrection the task of the Spirit is one of glorification. The sabbath becomes a motif for the future hope of creation which includes all of time and history. The glorious sabbath reflects the time of the new *aeon*, where linear historical time and biological cyclical time become caught up in the life of God. The new creation is radical in its newness and not simply a recreation of a lost paradise. While the eschatological tone of Moltmann's theology has the advantage of creating a vision for the future, its disadvantage is that at times his speculation seems to stray too far from the concrete biological issues which confront humanity in the ecological crisis. As a theological construct it may simply sound unbelievable to biologists.[104]

The advantage of Moltmann's theology is that it is both sensitive to a wide range of perspectives, yet avoids the somewhat bland approach of the World Council of Churches. His prolific writings on the subject have gone a long way to dispel the common image that theologians are not concerned about contemporary issues. His initial interest in political theology in his early *Theology of Hope* follows through in his theology of creation which has a thoroughly eschatological key. The advantage of a trinitarian approach is that each person of the social Trinity can shed a new insight on the relationship between God and creation. Overall, the focus of Moltmann's theology remains Christocentric. In his book the *Spirit of Life* it is clear that the prime interest of Moltmann's theology is the human community, rather than the community of the natural world.[105]

To sum up: the gathering momentum of the ecological crisis has led to a profound dis-ease in the human community. This anxiety is inevitable because of the growing consciousness of ourselves as historical beings who are separate from the natural world, yet who gather up the environment into our culture and history. There is a fear that something has gone wrong in our relationship with the natural world and a sense of powerlessness in the face of immense challenges. While most theologians do not go so far as to claim that theology can provide a solution, most agree that a change in attitude is necessary as part of the way forward. A religious perspective does not force its adherents to care for the earth, but it at least can ensure that it does not prevent a resolution. There is still a tendency to find scapegoats, either putting the blame on Christianity or a version of it such as Calvinism, or science and technology itself, or a mixture of both, or men, rather than women, or dualistic thinking as such. Such views need to be exposed as oversimplifications of the complex factors which led up to the ecological crisis. Martin Luther and John Calvin, for example, had a deep respect for the natural world and considered it God's gift for humanity. The advantage of the belief that the ecological crisis was a result in a simple increase in human activity is that no one is specifically blamed. The disadvantage is that it can lead to no one taking responsibility for the crisis.

Contemporary theologians have suggested different ways of relating God, humanity and creation as a first step in healing the rift with the natural world. For Barth the right relationship with creation comes as a natural consequence of faith in the Sovereign Lord who created all things good. For Rahner the discovery of the transcendence in humanity marks a healing of the split between nature and grace which has dogged scholastic theology. Theology's contribution to science is through making it aware of its fragmented human nature, that is a challenge to its claim to absolute authority. For Küng the God of the contemporary world is not immune to the sufferings of creation, but all are included in a God who is a God of the present and future hope. Teilhard de Chardin and process theology have some commonality in linking theological concepts with the process of evolution. The Christocentric approach of Teilhard becomes the emergent Spirit in God's consequent being of process thought. Both suffer from a somewhat over-optimistic view of the development of the world. Eco-feminist theology reminds us that the inherent dangers of sexual oppression are built into the fabric of modern societies and as such do serve to encourage exploitative attitudes, including that towards the earth. A reimaging of God as Mother, the feminine divine, the world as God's body and a shift towards pantheism all offer new interpretations which encourage a change in attitude to creation. An Orthodox perspective retains a sense of the otherness of God, while encouraging us to see all creation as participating in the life of the Creator. Such a vision reinstates the importance of humanity as one who offers up the creation to God for its redemption. Its disadvantage is that it can seem somewhat static in an eschatological sense. The cosmic purpose of creation through a Christological paradigm is highlighted in Moltmann's theology. While he brings eschatology back to the foreground of theology, his visionary approach at times seems to lose moorings with concrete issues.

NOTES

1. D. Cupitt, 'Nature and Culture' in *Humanity, Environment and God*, N. Spurway (ed.) (Blackwell, Oxford, 1993), pp. 33–45.
2. See especially R. Grove-White, 'Environment and Society: Some Reflections', *Environmental Politics* vol. 4, no. 4 (Winter 1995), pp. 265–75.
3. K. D. Kaufman, *The Theological Imagination* (Westminster Press, Philadelphia, 1981), p. 213.
4. L. White, 'The Historic Roots of Our Ecological Crisis', *Science* vol. 145 (1967), pp. 1203–7.
5. See, for example, J. Barr, 'Man and Nature: The Ecological Controversy and the Old Testament', *Bulletin of the John Ryland Library* vol. 55 (1972), pp. 9–32.
6. A. Primavesi, *From Apocalypse to Genesis: Ecology, Feminism and Christianity*, (Burns and Oates, London, 1991), pp. 36–43; R. Radford Ruether, *Gaia and God: An Ecofeminist Theology of Earth Healing* (SCM Press, London, 1993), pp. 173–201.
7. S. Clark, *How to Think About the Earth: Philosophical and Theological Models for Ecology* (Mowbray, London, 1993), p. 9.
8. H. P. Santmire, *The Travail of Nature* (Fortress Press, Philadelphia, 1985), pp. 124–5.
9. M. Luther, *Lectures on Genesis*, vol. 1, trans. G. V. Schlick in *Luther's Works*, J. Pelican (ed.) (St Louis, Concordia, 1955).
10. *Ibid.*, p. 39.
11. *Ibid.*, pp. 46–7.
12. *Ibid.*, p. 36.

13. M. Luther, *Selected Psalms*, vol. 12, trans. P. Jaroslav in *Luther's Works*, J. Pelican (ed.) (St Louis, Concordia, 1955), p. 119.
14. *Ibid.*, pp. 120–1.
15. L. Gilkey, *Reaping the Whirlwind: A Christian Interpretation of History* (Seabury Press, New York, 1976), p. 176.
16. *Ibid.*, pp. 177–8.
17. *Ibid.*, p. 186.
18. H. P. Santmire, *op. cit.*, p. 128.
19. K. Thomas, *Man and the Natural World: Changing Attitudes in England 1500–1800* (Penguin, London, 1983), pp. 33–5.
20. *Ibid.*, p. 42.
21. R. G. Collingwood, *The Idea of Nature* (Clarendon Press, Oxford, 1945), pp. 6–9.
22. *Ibid.*, p. 5.
23. R. Hooykaas, *Religion and the Rise of Modern Science* (Scottish Academic Press, Edinburgh, 1972), pp. 78–81.
24. C. J. Glacken, *Traces on the Rhodian Shore* (University of California Press, Berkeley, 1976), p. 500.
25. K. Thomas, *op. cit.*, p. 15.
26. L. Gilkey, *op. cit.*, pp. 189–91.
27. *Ibid.*, p. 191.
28. H. P. Santmire, *op. cit.*, p. 134.
29. *Ibid.*, pp. 134–5; see also R. G. Collingwood, *op. cit.*, p. 110.
30. H. P. Santmire, 'Studying the Doctrine of Creation: The Challenge', *Dialog*, 21, 1982, pp. 195–200.
31. R. G. Collingwood, *op. cit.*, p. 9.
32. *Ibid.*, p. 121.
33. *Ibid.*, p. 122.
34. Y. H. Vanderpool, 'Charles Darwin and Darwinism, a Naturalized World and a Brutalized Man?', in *Critical Issues in Modern Religion*, R. A. Johnson, E. Wallwork, C. Green, H. P. Santmire and Y. H. Vanderpool (eds) (Prentice Hall, Englewood Cliffs, 1973), pp. 72–113.
35. R. G. Collingwood, *op. cit.*, p. 136. Henri Bergson (1859–1941) marks the culmination of vitalistic thought where reality becomes life itself. Ironically, perhaps, his reaction against empiricism and reliance on intuition is against the trend of post-enlightenment science which formed the basis for Darwin's theory of evolution. See also R. C. Solomon, *Continental Philosophy since 1750: The Rise and Fall of the Self* (Oxford University Press, Oxford, 1988), p. 108.
36. H. Hartwell, *The Theology of Karl Barth: An Introduction* (Duckworth, London, 1964), pp. 3–7.
37. K. Barth, *Church Dogmatics, Volume 3/1*, trans. J. W. Edwards, O. Bussey and H. Knight; G. W. Bromiley and T. F. Torrance (eds) (T. and T. Clark, Edinburgh, 1958), pp. 1–15. See also G. Bromiley, *An Introduction to the Theology of Karl Barth* (T. and T. Clark, Edinburgh, 1979), pp. 109–55.
38. K. Barth, *op. cit.*, pp. 13–15.
39. *Ibid.*, pp. 15–22.
40. *Ibid.*, pp. 25–30.
41. *Ibid.*, pp. 50 ff. The initial acts of God in creation did not take place in a creaturely context like other historical events and so are 'non-historical' in the sense that they are prior to human and natural history, pp. 76–81.
42. *Ibid.*, pp. 176–81.
43. *Ibid.*, pp. 330–3.
44. *Ibid.*, p. 366.
45. K. Barth, 'No', in *Natural Theology*, trans. P. Fraenkel (The Centenary Press, London, 1946).
46. K. Barth, *Church Dogmatics, Volume 3/1*, *op. cit.*, pp. 371–7.
47. K. Barth, *Church Dogmatics, Volume 4/3*, trans. G. W. Bromiley; G. W. Bromiley and T. F. Torrance (eds) (T. and T. Clark, Edinburgh, 1961), pp. 139–40.
48. *Ibid.*, p. 143.
49. P. Teilhard de Chardin, *The Phenomenon of Man*, trans. B. Wall (Collins/Fontana, London, 1959), pp. 332–4.

50. P. Teilhard de Chardin, 'Christianity and Evolution: Suggestions for a New Theology', 1945 (unpublished) in *Christianity and Evolution*, trans. R. Hague (Collins, London, 1971), pp. 173–86. See also D. Gray, *A New Creation Story* (The American Teilhard Association for the Future of Man, Chambersberg, 1979), pp. 4–5.
51. K. H. Weger, *Karl Rahner: An Introduction to His Theology* (Burns and Oates, London, 1980), p. 87.
52. K. Rahner, *Theological Investigations: Volume 1*, trans. C. Ernst (Darton, Longman and Todd, London, 1965), p. 317.
53. *Ibid.*, p. 302.
54. K. Rahner, *The Christian Commitment: Volume 1*, trans. C. Hastings, 3rd edn (Sheed and Ward, London, 1970), p. 63.
55. *Ibid.*, p. 75.
56. *Ibid.*, p. 74.
57. *Ibid.*, p. 78.
58. K. Rahner, *Theological Investigations: Volume 13*, trans. D. Bourke (Darton, Longman and Todd, London, 1975), p. 95.
59. H. Küng, *Does God Exist?*, trans. E. Quinn (Collins, London, 1980), p. 115.
60. *Ibid.*, p. 123.
61. *Ibid.*, p. 125.
62. *Ibid.*, p. 642, see also pp. 182–3; 623–33.
63. *Ibid.*, pp. 648–55.
64. *Ibid.*, p. 655.
65. P. Gregorios, *The Human Presence: An Orthodox View of Nature* (World Council of Churches, Geneva, 1988), p. 57. The difference between God and creation according to Orthodox tradition is also explained in V. Lossky, *Orthodox Theology: An Introduction*, trans. I. Kesarcardi-Watson (St Vladimir's Seminary Press, New York, 1984), pp. 51–78.
66. V. L. Lossky, *The Mystical Theology of the Eastern Church*, trans. the Fellowship of St Alban and St Sergius (J. Clarke, London, 1957), pp. 79–80.
67. S. Bulgakov, *Sophia: The Wisdom of God: An Outline of Sophiology* (Lindisfarne Press, Hudson, 1993).
68. J. D. Zizioulas, 'Preserving God's Creation: Three Lectures on Theology and Ecology', II, vol. 12, no. 2, *Kings Theological Review* (1989), pp. 41–5.
69. *Ibid.*, p. 45.
70. J. D. Zizioulas, 'Preserving God's Creation: Three Lectures on Theology and Ecology', III, vol. 13, no. 1, *Kings Theological Review* (1990), pp. 1–5.
71. *Ibid.*, p. 2.
72. *Ibid.*, p. 5.
73. V. Rossi, 'The Earth is the Lord's', *Epiphany*, vol. 6, no.1 (1985), pp. 3–6.
74. V. Rossi, 'Theocentrism: The Cornerstone of Christian Ecology', *Epiphany*, vol. 6, no. 1 (1985), pp. 8–14.
75. B. Meland, 'Introduction' in *The Future of Empirical Theology*, B. Meland (ed.) (University of Chicago, Chicago, 1969), pp. 1–62.
76. J. Cobb, *A Christian Natural Theology* (Lutterworth, London, 1965), pp. 31–4.
77. H. Smith and S. Todes, 'Empiricism: Scientific and Religious' in B. Meland (ed.) *op. cit.*, pp. 129–146 and J. Cobb, *Is it Too Late?* (Beverley Hills, Bruce, 1972).
78. J. Cobb, *op. cit.*, pp. 51–2.
79. *Ibid.*, p. 111.
80. As I argued in the first chapter process theologians seem to be arguing for a re-enchantment of science back to the early beginnings of science where a belief in the supernatural in nature did not hinder technological advancement. See *The Re-enchantment of Science*, D. Griffin, (ed.) (State University of New York Press, New York, 1988).
81. R. Radford Ruether, *New Woman: New Earth* (Seabury Press, New York, 1975), p. 80.
82. *Ibid.*, p. 195. The link that Ruether makes between patriarchy and a dualism of mind and matter seems rather exaggerated. Her more recent book seems clearer as she draws a more general parallel between all systems of dominance. See R. Radford Ruether (1993), *op. cit.*, pp. 173–201.
83. She is attracted by the ecological spirituality which seeks to worship Gaia, rather than God. However, she is critical of feminist worship of the goddess and seeks to move beyond this in linking the divine God with the 'life power from which the universe arises', R. Radford

Ruether (1993), *op. cit.*, pp. 4–5. For a review see C. Deane-Drummond, *Theology in Green*, vol. 8 (1993), pp. 40–2.
84. R. Radford Ruether, *ibid.*, p. 1.
85. Ibid., p. 249.
86. D. Soelle and S. Cloyes, *To Work and to Love: A Theology of Creation* (Fortress, Philadelphia, 1984); U. King, *Women and Spirituality* (Macmillan, Basingstoke, 1989).
87. D. Soelle and S. Cloyes, *op. cit.*, p. 5.
88. *Ibid.*, p. 222.
89. E. Moltmann-Wendel, *A Land Flowing with Milk and Honey*, trans. J. Bowden (SCM Press, London, 1986), p. 98.
90. *Ibid.*, p. 160.
91. C. Lasch, *The Minimal Self* (Pan/Picador, London, 1985), pp. 246–53.
92. S. McFague, *The Body of God* (SCM Press, London, 1993).
93. *Ibid.*, p. 149.
94. *Justice, Peace and the Integrity of Creation*, G. Limouris (ed.) (World Council of Churches, Geneva, 1990); see also D. Gosling, *Justice, Peace and the Integrity of Creation* (Council of Churches for Britain and Ireland (CCCBI), London, 1993).
95. C. Birch, 'Nature, God and Humanity in Ecological Perspective', *Christianity and Crisis*, vol. 39 (1979), pp. 259–66; H. Sindima, 'Community of Life', *The Ecumenical Review*, vol. 41, no. 4, (1989), pp. 537–51.
96. J. C. Peck and J. Gallo, 'JPIC: A Critique from a Feminist Perspective', *The Ecumenical Review*, vol. 41, no. 4, (1989), pp. 573–81.
97. R. Preston, 'Humanity, Nature and the Integrity of Creation', *The Ecumenical Review*, vol. 41, no. 4 (1989), pp. 552–63.
98. J. Moltmann, *God in Creation: An Ecological Doctrine of Creation*, trans. M. Kohl (SCM Press, London, 1985). For a detailed commentary of Moltmann's theology of creation, see C. Deane-Drummond, *Towards a Green Theology through Analysis of the Ecological Motif in Jürgen Moltmann's Doctrine of Creation* (Manchester University PhD thesis, 1992); for a summary see, C. Deane-Drummond, 'A Critique of Jürgen Moltmann's Green Theology', *New Blackfriars*, vol. 73 (1992), pp. 554–65.
99. J. Moltmann, *The Trinity and the Kingdom of God*, trans. M. Kohl (SCM Press, London, 1985), p. 19.
100. J. Moltmann, *God in Creation*, p. 88. He links the divine withdrawal with the feminine in God so that 'In a more profound sense he "creates" by letting-be, by making room and by withdrawing himself. The creative making is expressed in masculine metaphors. But the creative letting-be is better brought out through motherly categories.'
101. J. Moltmann, *Theology of Hope: On the Ground and Implications of a Christian Eschatology*, trans. J. W. Leitch (SCM Press, London, 1967), p. 282.
102. J. Moltmann, 'Human Rights: The Rights of Humanity and the Rights of Nature' in *The Ethics of World Religions and Human Rights*, H. Küng and J. Moltmann (eds) (SCM Press, London, 1990), pp. 120–35.
103. J. Moltmann, *Theology of Hope*, p. 137; see also J. Moltmann, *The Way of Jesus Christ*, trans. M. Kohl (SCM Press, London, 1990), pp. 250–63. Moltmann comments in the latter book that Christ's 'resurrection is indeed the beginning of the new creation in which death will be no more . . . This transformation is its eternal healing', p. 258.
104. See C. Deane-Drummond, *Towards a Green Theology*, pp. 286–322.
105. J. Moltmann, *The Spirit of Life*, trans. M. Kohl (SCM Press, London, 1992).

3

Environmental ethics

While the philosophy of nature is certainly not anything new, dating back to pre-Socrates, it is only recently that modern philosophers have concentrated their energies on this issue in the light of contemporary environmental concern. In the past, mainstream moral, political and social philosophy has focused on the social environment, to the exclusion of the natural environment. Ideas and values stemming from environmentalism and conservationism have been greeted with suspicion by mainstream Western metaphysics, epistemology and ethics.[1] Nonetheless, in the last twenty years or so there has been a resurgence of interest in environmental questions. Andrew Light and Eric Katz, who are leading philosophers from Canada and the USA, have aptly stated:

> The problematic situation of environmental ethics greatly troubles us, both as philosophers and as citizens. We are deeply concerned about the precarious state of the natural world, the environmental threat that threatens humans, and the maintenance of long-term sustainable life on this planet. The environmental crisis that surrounds us is a fact of experience. It is thus imperative that environmental philosophy, as a discipline, address this crisis – its meaning, its causes and its possible resolution.[2]

While it is generally acknowledged that philosophers from North America have been most active in this field, in Britain too, there has been a steady growth in research activity, including a conference in 1986 for the

Society of Applied Philosophy on environmental and welfare themes, a journal edited at Lancaster University entitled *Environmental Values*, and in 1993 the Royal Institute of Philosophy held a conference to consider the themes of nature, the natural environment and the related issues of values, ethics and society, to name but a few examples. I intend to outline some of the basic conclusions emerging from these studies, to indicate the internal debates and critique from philosophical ethicists, as well as point to areas where theological analysis can make a contribution to the debate. I will leave a detailed study of biotechnology as a specific case for theological and philosophical reflection until the next chapter and I will also be dealing with specific ethical and philosophical issues around the question of sustainability later in this book.

Who/what has moral standing?

The traditional domain of ethicists has been the consideration of the moral responsibility human beings have for each other, either as individuals or as social units. The idea that the environment might be an issue for ethics, seems at first sight rather surprising. The most conservative approach we could adopt is 'How do environmental issues affect human interests?' In other words, are there specific environmental problems which lead to conflicts of interests within the human community? The philosopher John Passmore wrote an influential book in the 1970s entitled *Man's Responsibility for Nature*.[3] He insisted that a concentration on human interests, or anthropocentrism, as long as it showed some responsibility for care of the natural environment, was perfectly sufficient. Other philosophers were far less assured that such a focus was in any way adequate to meet the perceived dangers, which seemed to be of unparalleled proportions.

There are various possible strategies. One is to extend moral concern beyond the human community to animals and other sentient beings. In this case, the ability to feel pleasure or pain, as well as the psychological categories of emotion and relationship, are extended to include animals. Paul Singer and Tom Regan are leading advocates of this position.[4] It has been used as a theoretical basis for environmental rights activism which I will return to later in this chapter. It is also possible to extend moral concern to all living things, not just humans and animals. The criteria for moral standing is now being alive, rather than the ability to feel pleasure or pain. This idea took its inspiration from the life and work of Albert Schweitzer. He began his career with a doctorate in theology and philosophy, but subsequently trained in medical science and became a missionary doctor in French equatorial Africa in 1913. This remarkable man believed that we should show a 'reverence for life', the *Ehrfurcht* or sense of awe and mystery when faced with the natural world.[5] He insisted that the will to live is in every living being. He championed the animal protection movement. Paul Taylor has written an influential book which draws on Schweitzer's ideas as a basis for environmental ethics entitled *Respect for Nature: A Theory of Environmental Ethics*.[6] Once we extend moral consideration to include all living things we have a biocentric, rather than anthropocentric outlook on nature. Taylor believes that we should consider each being as having a good of its own, without relating this to human subjective value concepts. Taylor develops the idea of inherent worth. According to his definition:

> Inherent worth is not the same as the good of a being. Inherent worth affects the way we behave towards a living thing, whereas we can say that they all have good of their own. If an entity has inherent worth it is worthy of respect.[7]

Unfortunately, the terminology used by ethicists can be confusing, as different terms have different meanings depending on the particular locus of value along the scale of anthropocentrism to biocentrism. In general, intrinsic value refers to the positive value of an event or condition in and of itself. For Taylor, positive value is associated with enjoyment. If something is a means for further ends it is said to have instrumental value. Now while anthropocentric philosophers will give living things instrumental value, in as much as they are of direct usefulness or benefit for human beings, they are more reluctant to give living things intrinsic value. Inherent value is a slightly more difficult term to understand. It generally refers to the value given to an object by the valuing subject. Whereas intrinsic value shows a value independent of the observer, for inherent value the subject must be present. People will give values to certain places of beauty, cultural or historic interest, but this is a subjective assessment based on current cultural norms. Taylor uses inherent worth to stress the way we *behave* towards living things, they are worthy of respectful behaviour, regardless of our subjective sense of inherent value. That is, our behaviour should seek to preserve an entity's good, independent of our own ranking according to benefits to humans. For Taylor, respect for nature is an ultimate one; 'it is itself one of the most fundamental kind of moral commitment that one can make'.[8] The basis for this commitment stems from what Taylor calls the 'biocentric outlook on Nature'. There are four key beliefs which form the basis for this philosophy:

1 human beings are members of the earth's community of life in the same sense and on the same terms as other living things;
2 human beings are integral elements in a system of interdependence, such that the survival of each living thing is determined both by physical conditions and relations to other living things;
3 all organisms are teleological centres of life, such that each has the possibility of pursuing its own good in its own way;
4 humans are not inherently superior to other living things.

Does this notion remove any ideas of difference? Taylor suggests that while we acknowledge difference, it is 'put aside' in order to stress our common origin and common environment. We need to enter imaginatively into the world of another organism and see it from this perspective. This is different from 'reading in' human consciousness, which is the transfer of human emotions to animals, etc. Rather, it is independent of human interests. Once we have entered imaginatively into this perspective it allows us to come to a moral commitment based on respect. This view acknowledges our recent arrival on the planet and the hard fact that other species can do without us, but this does not apply in reverse. He believes that to insist that we are in any sense the culmination of evolution is simply an expression of human vanity. For Taylor, the earth would actually benefit from human extinction; 'our presence, in short, is not needed'.[9] It seems to me that Taylor's argument is pulling us in two different directions. On the one hand, we are encouraged to commit ourselves to all living things on this planet from their perspective, but on the other hand we are not deemed worthy of respect ourselves. Moreover, while his theory seems to argue for equality between humans and all living things, in practice he allows for the survival interests of humans to take precedence over the survival of life.[10]

The biocentric view does not entail placing value on the earth as a whole, though it does seem to lead in this direction in spite of Taylor's remonstrations. Once we do place value on the earth as a whole, or even on a species as a whole, the centre of value finds a new locus in the whole, or holism, rather than biocentrism. Holmes Rolston III is one of the most influential philosophers who have adopted this approach. He acknowledges that an ecosystem has 'no genome, no brain, no self identification . . . But to find such characteristics missing and then to judge that ecosystems do not count morally is to make another category mistake.'[11] An ecosystem does have a history of development, but the laws which govern such changes are according to mathematical statistics, rather than genetics. He believes that humans have duties at the level of 'survival units'; the organism is one level and the ecosystem is another. He argues against automatically giving priority of community over individual, which softens his view compared with some so-called 'deep green' philosophers such as Arne Naess. He has coined the idea of *systemic* value, that is placing value on systems as a whole, in addition to their components. He believes that this new term is necessary and instrumental value and intrinsic value have limited application. Hence:

> We are no longer confronting instrumental value, as though the system were of value instrumentally as a fountain of life. Nor is the question one of intrinsic value, as though the system defended some unified form of life for itself. We have reached something for which we need a third term: systemic value. This cardinal value, like the history, is not at all encapsulated in individuals, it too is smeared out into the system. The value in this system is not just the sum of the part values. No part values increase of kinds, but the system promotes such increase. Systemic value is the productive process; its products are intrinsic values woven into instrumental relationships.[12]

Rolston still wishes to acknowledge the value of humans as having the 'highest value attained in the system'; but he insists that even the most valuable of the parts is less than the whole. His main concern in more recent writing seems to be to shift our attention away from simply human interests. Responding to the claim that we commit the 'naturalistic fallacy' in finding value in nature, he argues that:

> Rather the danger is the other way round. We commit the subjectivist fallacy if we think that all values lie in subjective experience, and worse still, the anthropocentrist fallacy if we think that all values lie in human options and preferences.[13]

In this way it makes it easier for him to differentiate between human and non-human interests, compared with Taylor's model.

Nonetheless, there are practical difficulties associated with the 'overriding value' in the systemic process. Will any environmental policy maker place greater value on a system which seems ill-defined and itself is in a state of flux? It seems to me that the biological basis of Rolston's ideas is naive. His views tend to reinforce the concepts of integrity and stability in ecosystems in ways which do not exist in practice. While it is true that the idea of placing value on wholes is an important corrective to more isolated individual bases for ethical consideration, there is a strong resistance to what can be seen to be a form of 'ethical fascism'. He argues for extending value to the whole earth as well, such that 'environmental valuing is not over until we have risen to the planetary level'.[14] For him the earth as a whole has importance as the ultimate source of all life and all value. 'The creativity within the natural system we inherit, and the values this generates, are the ground of our being, not just the ground under our feet. Earth could be the ultimate object of duty, short of God, if God exists.'[15] He argues against the idea that there can be no value without a valuer, that there has to be some kind of consciousness of being in order to possess value. Values, then, go beyond our constructions.

Gaia as an ethical basis for value?

Holmes Rolston III seems to be coming close to the view that takes the whole planet as a guide for moral value. The Gaia hypothesis, originally devised by James Lovelock, examines the reasons for the apparent stability of environmental conditions in the biosphere, that is the envelope of gases surrounding the earth. He concluded from his research that living organisms together keep environmental conditions constant. This is in contrast to the traditional view that life simply adapts to external conditions.[16]

Scientists do not agree as to how the Gaia hypothesis might function. The lowest common denominator is that there are interactions between living things and the environment, so that looking at the system as a whole is an important subject for research. A more organismic view of the earth is more controversial. However, this idea is not new. As early as 1785, the Scottish scientist James Hutton (1726–97) supported the idea that the earth functioned as an organism, as did the Russian scientist Vladimir Vernadsky (1863–1945) in the last century. The belief that the whole planet cooperates as a single system seems to suggest a cooperative model for evolution, rather than the more competitive model of Charles Darwin. This cooperative model slides into an ideological and teleological model, where Gaia becomes part of a philosophical quest for life.[17] While Lovelock has specifically denied that Gaia is teleological, some of the language he uses opens up such an interpretation. He admits that we cannot prove that Gaia is 'alive', but to attempt to do so is 'otiose'. The direction of Gaia remains the survival of life. This aspect of Gaia is ignored by some philosophers taking Gaia as an inspiration for cooperative models between humanity and nature.

An understanding of Gaia which includes homeostatic processes can lead in two completely different directions. One approach, which I have indicated here, is through links with systemic value. I will elaborate on this approach first before discussion of the alternative resource management approach. The idea of collective value is certainly not new and stems from the pioneering work of Aldo Leopold, whose book *A Sand County Almanac*, written in 1949, set down the paving stones for what was to become the basis for models of environmental ethics.[18] He argued that the land, that is soils, waters, plants and animals, all deserve moral consideration as part of the community of life.

It is possible to adopt Leopold's views while remaining convinced that human beings generate these values. Baird Callicott takes this approach, so that human beings have evolved in such a way that they treat non-humans as part of their community.[19] His sociobiological approach fails to distinguish adequately between human behaviour that arises from cultural influences and that which arises from biological determinants. The idea that Gaia acts as a model for altruism seems to be more innocuous and less deterministic. However, Sahouris suggests that we are subject to the workings of the planet.[20] Unless humans become in tune with the workings of the planet we will fail to make progress. But is the scale of value suggested by Gaia so obvious? If we follow Lovelock's ideas to their logical conclusion, the most respect should be given to the micro-organisms on the earth, which are the main contributors to environmental stability.

This logic would suggest that human beings are mere parasites on the planet.[21] This bears some resemblance to Taylor's suggestion, from a biocentric perspective, that human beings are no longer necessary for life on earth. However, the protagonists of Gaia are not as concerned as Taylor to spell out their ideas in ethical terms. This is similar to deep ecology's focus as a consciousness movement, rather than an ethic as such.[22] Pedler, for example, speaks of a new lifestyle that is Gaian, where 'the human race is an integral part of a single life force, sometimes called the earth organism, the earth spirit of Gaia'.[23]

It is this extension of self into a new consciousness which is highly problematic and has led to charges of ecofascism. An extension of the self into the world tends, ironically, to project anthropocentric ideas into the world, which is the opposite of the dream of 'deep ecology'. While Holmes Rolston III is careful to avoid any ideas of extension of the self, it seems to me that once we try to identify value outside ourselves it becomes all too easy to link our consciousness with that perceived elsewhere. In other words, the empathy necessary may be idealistic, in practice such a view leads to identification. Once this takes place anthropocentrism returns.

John Milbank has reached a similar conclusion from a rather different perspective.[24] He argues that a concern for nature in itself separates subject from object. Once we treat nature as an object it becomes part of the project of modernity. Any 'turn to nature' is unlikely to be the locus of 'value', as thinking of nature in this way as separate from humanity is itself a modern concept. I have indicated that we cannot help but think of nature in anthropocentric ways, for all the rhetoric on biocentrism and holism.

The ambiguity in Gaia as a basis for ethics becomes more obvious in those who argue that their views on the earth as a resource has its basis in Lovelock's hypothesis. These views are unashamedly anthropocentric where action is judged according to the possible benefit or otherwise to human beings. John Passmore, as I noted above, insists that humanity alone generates values and there are no proper grounds to base value outside the human community. If we adopt a view of Gaia that requires us to make the least number of assumptions, we could argue that the ability of the earth to correct itself after change shows the robustness of our planet. Hence, human beings are free to do what they like, all talk of impending doom seems unrealistic and unfounded. Alternatively we could take steps to protect just those areas of the earth which act like the 'vital organs', namely the rainforests and oceans. Lovelock himself seems to reject any resource management ideas.[25] He believes that it is unrealistic and arrogant to believe that we can manage the earth, when we have failed so dismally to manage human affairs. Instead, we should act as 'representatives' of other life forms on the planet, living in partnership and respect for all life.

The case for animal rights?

Those who argue for an ethical basis for animal rights take a very different view from that outlined above. An individual, rather than collective, approach seems to be the norm. The appeal is often to those characteristics in animals which arouse most sympathy with human kind: rabbits and dogs, whales and dolphins are championed in campaigns, but little attention is given to humbler life forms under threat, such as insects or amphibians, even though these creatures may be just as important to the ecosystem. In addition, while those who are known as 'deep ecologists' may consider themselves to have subversive tendencies, theirs has characteristically been a quieter revolution. More recently this has begun to change, with militant protesters campaigning against, for example, the Newbury bypass in late 1995 and 1996.

The more established and militant animal rights activists have put the issue of our treatment of animals and the ethics of some scientific practices firmly on the map of matters which require urgent attention. This may be one reason why, in general, the concern for the environment as a whole seems to be losing some of its momentum compared with five years ago. The global problems we face just seem too vast and too complicated to contemplate. Moreover, we have been told so many times that we will die in an ecoholocaust that we are beginning to become sceptical of the experts. Our lifestyles seem to carry on the same as ever, our resilience reinforced by other panics such as those over nuclear disaster and meteor strikes that come and go. We do not necessarily believe in a crisis, or at least we are not prepared to adjust our lifestyles even if we do believe. As Stephen Clark has aptly pointed out:

> . . . our willingness to sacrifice our own present interests for the general good of humankind or of the world is feeble; advising others not to cut down rainforests or pollute the seas is easy enough; preserving our own wetlands or cutting back on burgers or not travelling to conferences on conservation by convenient means is another matter.[26]

Compared with these general environmental issues those connected with animal rights seem to be more obvious and more attainable. Moreover, we are reminded of the issue by animal rights protesters. Even if we do not agree with this form of behaviour, we are not allowed to forget.

Paul Singer and Tom Regan have actively worked for the establishment of animal rights as part of an environmental ethic. Both are committed to an appreciation of the individual worth of animals, as opposed to what Singer describes as unwarranted 'speciesism', that is concern just for our human species.[27] Singer is convinced that as moral agents we have a duty not to cause suffering to sentient beings, which includes humans and animals. In other words, he uses the criterion of sentience as a basis for moral worth. Singer argues against use of animals in experimental research, on the basis that as animals feel pain our treatment of them should be the same as our treatment of humans. He believes that just because animals are unlike humans in many respects, this is no excuse for not treating them in a moral way and respecting the interests of the creature. It is clearly in the creature's interest to avoid pain.

Tom Regan sets out his case for animals rights in his book, *The Case for Animal Rights*.[28] He argues that it is fundamentally wrong to treat animals as a resource. The animal rights movement has three main goals:

1. to abolish the use of animals in science;
2. to ban commercial animal agriculture;
3. to eliminate commercial and sport hunting and trapping.

He writes with passion and enthusiasm when he claims that:

> the whole creation groans under the weight of the evil we humans visit upon these mute, powerless creatures . . . All great movements, it is written, go through three stages: ridicule, discussion, adoption. It is the realisation of this third stage – adoption – that demands both our passion and our discipline, our heart and our hand. The fate of animals is in our hands. God grant that we are equal to this task.[29]

We might now ask: how are we to distinguish between the choice of different species? Do all animals have rights? Should some be given priority over others? Van de Veer addresses this issue in his schematic analysis of positions we could adopt, ranging from what he terms 'Radical Speciesism', which gives no interest whatsoever to the non-human, to 'Species Egalitarianism', where all species are treated equally.[30] In between these extremes he posits the notion of 'Extreme Speciesism', where the basic interests of animals, that is survival interests, are subordinated to the peripheral interests of humans. In this scenario we could quite happily kill mink, for example, for fur coats as a luxury item. Where there is no conflict of interests this view might allow us to promote the interests of the animal. For example we could argue for the protection of a species where this was not to the detriment of humans. However, it is hard to think of many practical situations where this would apply.

For 'Interest Sensitive Speciesism', the basic interest of an animal takes precedence over the peripheral interests of humans, while the basic interest of humans takes precedence over the basic interests of animals. In this scenario humans could kill animals for survival needs, but not for peripheral interests. If we apply the example above, all killing of minks for fur coats would be outlawed. The difficulty with this view is that the differences between species are ignored. In order to combat this de Veer has come up with his own theory of 'Two Factor Egalitarianism'. In this case a hierarchy is introduced such that the most psychologically developed species takes precedence over the least psychologically developed. Such a scheme would not, for example, allow us to sacrifice a chimpanzee for its kidney in order to help a child with Tay Sachs disease. In other words 'being a member of *Homo sapiens per se* is not assumed to justify preferential treatment of humans over animals'.[31] Van de Veer is well aware of the potential problems for two factor egalitarianism. For example, how do we discern what is basic interest and what is peripheral? How do we determine levels of psychological activity? What about intraspecific justice? More important, perhaps, how do we enforce such a view? What about the significance of wholes and communities? This view is still thoroughly individualistic and takes its bearing from anthropocentric concerns.

Lawrence Johnson, in his book *A Morally Deep World*, draws on Singer's views, but seeks to develop them further.[32] He uses overall life and well-being as a criterion for interests, rather than simple pleasure or pain. He also takes a less individualistic line compared with Singer and opts for moral consideration of a 'gene lineage', so that the whole has a greater moral significance than the sum of the parts. He argues that we do not need to choose between the animal rights views of Singer and Regan and holistic values, rather 'we must have it both ways'.[33] His concept is a clever one, which might seem to avoid the charge of fascism of holistic 'deep' schemes and individualism and anthropocentrism of so-called 'shallow' schemes, which focus on individual rights of humans and by extension to animals. However, I am less convinced that it would really work in practice.

Stephen Clark also argues strongly in defence of animals. He makes the important point that our preferential treatment of them is more to do with our particular symbolic needs, rather than realistic ones:

> Those who mourn a dead rabbit are utterly indifferent to a dead mouse; those who would not kick a dog readily support the torture of the equally intelligent pig. These discriminations may sometimes reveal a total indifference to animals – as the experimenter who consoles objectors by remarking that he looks on experimental animals and people's pets in utterly different ways, thereby showing that he sees the only possible evil involved as an offense to the sentimentality of pet owners. But this indifference feeds upon our fantasy – of cows as placid and incurious hunks of flesh, of dogs as loyal slaves, of birds in general as so much stuffed and twittering fowl. We do not see the individuals; we see types of our imagination.[34]

According to Clark it is these fantasy projections on animals which seem to act as a source of resistance in any adoption of animal rights. Yet it seems to me that the ability to feel anything for animals, through pet ownership, for example, can make us more sensitive to animals in general and generate the passion needed for resistance. Is vegetarianism the only option? Stephen Clark believes that it is, even if we adopt the minimal principle of debarring unnecessary suffering to animals. He argues that as 'flesh eating' is a luxury over and above our survival needs, we cannot justify it.

Andrew Linzey takes a similar line, though he writes from a more explicit theological perspective.[35] It seems to me that while the ideal of vegetarianism is one worth aiming for, the impression Linzey, and to some extent Clark, give in some of their writing is an overly judgemental attitude to meat eaters. The 'vampire' language that Linzey uses is alienating to anyone who is not vegetarian. A more realistic goal would be to work for a greater awareness of the suffering of animals in much farming practice and a more moderate intake of animal products, rather than a total ban. It seems to me that the survival needs of farmers and all those associated with farming have to be taken into account, along with animal rights.

Linzey also asks the knotty question of how far we are justified in extending the language of 'rights' to animals. Can animals have 'rights' without 'responsibilities'? Linzey argues that the criteria of personhood and rationality are far too restricting as criteria for animal rights. Moreover, the ultimate giver from a theocentric perspective is God, who pronounces creation good independently of humans. Whatever the theoretical arguments for and against the use of animal 'rights', it seems to me that it is a phrase that is here to stay and one which has encouraged the ground swell of opinion in support of animal welfare.

Linzey's theological approach is interesting as it enables additional support for humane treatment of animals beyond the traditional 'stewardship' approach. He acknowledges that the eschatological future is one where human beings and animals live in harmonious relationship. He draws on Colossians in order to support a cosmic Christology which extends redemption and liberation to include animals. My question now would be: why stop here? Once we accept that this passage refers to the cosmos, rather than narrowly to human beings, why not include all of creation as well? Is Christ the liberator not just of animals, but all creatures? While his theological arguments are consistent with his views, I am less convinced that he has shown that God's concern has a priority for animals compared with all life. It is true that if we use sentience as a criterion for liberation, then animals and humans fall easily into the same category. It would then be natural to restrict liberation to animals. However, I am more of the opinion that sentience alone is inadequate in a construction of environmental ethics, as it fails to serve the wider needs of communities, ecosystems and other life forms.

Sentience is, in spite of the rhetoric, anthropocentric in its orientation. While it may be less objectionable than rationality as a basis for moral concern, it seems to me to ignore the whole complexity of interrelationships. Furthermore, the language of rights joins in with the chorus of other rights movements: the rights of women, the rights of blacks, the rights of all marginalized groups. As such, it tends towards a specialism and splintering effect on the human community. It is interesting that feminist theologians are now moving away from a campaign of equality to one which acknowledges difference.[36] It is this difference between men and women that can lead to a deeper sense of respect for each other. Is it not time that we acknowledge difference between humanity and the world, while at the same time working for greater cooperation? Stephen Clark comes to a similar conclusion in his essay on global religion:

> If we are to cope with our crisis, we must recognise the World as other than the human world, and recognise ourselves as inextricably dependent on that World. It is both our Other and our Origin, something unconstrained by our projected values and recognised as something by which we should be constrained. . . . Moralists have tended to suggest that it is insofar as things are like 'us' that they are deserving of respect: but the better way is to respect them as not being ourselves, and so allow them to *be* . . . Of course, since we are utterly dependent on the world (and so a minor part of it) there can be no gap between Us and It. But what It is does not depend on what we say it is. All attempts to evade this fact, like similar attempts to evade the laws of logic, seem to me to be appallingly misguided.[37]

The end of anthropocentrism?

Does the logical conclusion of these deliberations mean that we can no longer sensibly speak of the centrality of human beings in the cosmos? The philosopher Mary Midgley suggests that it is inevitable that we look at the world in some sense from a human viewpoint.[38] Our powers of sympathy are necessarily limited. Moreover, self-love is a prerequisite for love of other creatures. This is distinct from the selfish independence that ignores the other. The two extremes of viewing humanity as the 'crown of creation' or the 'parasites on the planet' sit uneasily alongside each other. Midgley suggests that there is a right sense in which humans need to have a particular concern for their own species. She argues, further, that:

> From a practical angle, this recognition does not harm green causes, because the measures needed today to save the human race are, by and

> large, the same measures that are needed to save the rest of the biosphere. There is simply no lifeboat option by which human beings can save themselves alone, either as a whole, or in particular areas.[39]

The anti-anthropocentric rhetoric more commonly refers, in practice, to a narrowly conceived focus on human interest alone, so that 'The kind of anthropolatory that would always set human interests above other life forms is surely no longer defensible'.[40] I agree with the main thrust of her argument, which also seems to be more realistic in its assessment of the current situation.

It is, indeed, a focus on the practical aspects of environmental ethics that have stopped current philosophers short in their tracks. How far can any of these theoretical debates about value and so on apply to practical policy making? Alongside this doubt there is a strong stream in current philosophy which resists all attempts to lay down 'foundations' as a starting point for discussion.[41] Nonetheless, this does not mean that philosophical ethics have no theory at all; rather the move is towards the philosophical pragmatism of John Dewey, Charles S. Peirce, William James, George Herbert Mead and Josiah Royce.[42] A synthesis between the thought of these authors, writing in the early part of the century, and current environmental concern leads to a mode of thinking known as 'environmental pragmatism'.[43]

According to philosophical pragmatism there is no longer an absolute distinction between subject and object, mind and the world, knower and known. While some of the pragmatists were openly hostile to metaphysics, all agreed that their analysis needed to make sense of experience and not overstep the limits of knowledge imposed by it. The outcome of this view is, above all, to see the world in all its diversity. Reality becomes process and development. The rightness of an action now becomes 'system'-dependent. The means which seem to serve us best is the only criterion. Antony Weston uses the analogy of ethics as making our way furtively through a 'swamp', rather than building a pyramid on a bedrock foundation. What is needed tomorrow is not the same as today, the world of values is continuously shifting and changing.[44]

The practical outcome of this view is that it leads to moral pluralism. However, it is not necessarily associated with metaphysical pluralism, where we play 'musical chairs' with our basic assumptions.[45] A modified view is a single metaphysic that acknowledges pluralism in the world. Kelly Parker suggests that a pragmatist would accept anthropocentric, biocentric and ecocentric as all having a valid place.[46] Sometimes we might focus on sustainability of the whole, at other times on the individual right. While this may apply in certain cases, this idea does little to suggest ways forward when there is a conflict of interest between these two scenarios. It is always human experience which is the locus of interest for pragmatists, so I am less convinced that such a view allows for the more holistic and biocentric positions. Kelly also suggests that according to pragmatism the difference between instrumental and intrinsic value ceases to be important, rather all is now seen as part of a web, intricately interconnected with the other in relation. I am worried by this idea as it seems to prevent us from making adequate distinctions between creatures. If we no longer have any sense of instrumental value, then the logic is that all objects, even inanimate ones, become subject to our concern as part of our experience. It seems to me that environmental pragmatism is, at this level, profoundly disappointing. It seems to convey the notion of a 'down to earth' approach to environmental issues. However, far from being practical, it seems to remove any criteria at all from the equation. I would agree that practical issues need to be taken into account. However, I have my doubts about the potential of philosophical pragmatism to give us guidelines.

While I would agree with the general resistance in environmental pragmatism to a crude moral monism, there is a place for some theoretical foundations to give us bearings in the 'swamp'. Philosophical environmental pragmatism as such, drawing on the American philosophical tradition, seems to fail in this respect. Andrew Light has a good point when he suggests that much of the literature on environmental ethics has become intolerant and vindictive, the very opposite of what is needed in order to work together for the solution of practical environmental problems.[47] He also suggests that metaphilosophical environmental pragmatism 'is not a dogmatic pluralism, committed to some version of post modern relativism which admits no possibility for moral realism or foundationalism. It is metaphilosophical, and thus not necessarily closed to the idea of formulating a rich foundationalism in environmental ethics.'[48] I agree with Light that the adoption of any single position as the solution to the complex problems of environmental ethics is bound to fail. The difficulties of philosophical environmental pragmatism based on pragmatist philosophy are similar to those of process philosophy and theology, which I will discuss in more detail below.

A theological critique

In general, the debates within environmental ethics take place from within the boundaries of secular philosophy. There are some who remain deeply sceptical about the value of any theological contribution to the debate.

Frankena, for example, raises the possibility of theocentric ethics, but rejects this approach on the grounds that he finds it 'troubling' to make ethics dependent on theology.[49] His unease is most likely related to the distrust amongst philosophers of any approach which seems to be 'dogmatic', laying down prior foundations. I intend to show in this section that the rejection of theology *per se* is naive; in fact a rejection of spirituality and awareness of the sacred in the natural world has contributed to the mechanistic approach which is at the heart of the environmental crisis.

Some leading philosophers are aware of the potential usefulness of spirituality, yet it still remains vague and an often hazy outline that lacks theological depth. Holmes Rolston III, for example, makes an impassioned plea for giving value to the wilderness in his book *Philosophy Gone Wild*.[50] As I mentioned earlier, he reaches out for greater universalizing in our thinking, so that in other works he speaks of systemic value, that is value to systems. For him nature is a source of values, a generative process with which we connect and find identity. He restricts his understanding of 'nature' to the earth, rather than being cosmic in scope. From this he invites his readers to 'take the experiential plunge into nature, mixing participatory immediacy and reflective distance, reason and emotion, romance and criticism, nature and spirit . . . We go out into the field. We go wild.'[51] What he seems to be saying is that we need a reconnection with all that is in nature, including a primitive instinct for spirituality, a pantheistic or animistic mode of being. My suspicion seems supported by his comment on the contribution of Christianity. For him Christianity:

> did not so much replace, as complement, enrich and extend the primitive and universal impulse in us to celebrate the return of the warmth of spring and the resurgence of life that is given by these mysterious powers of the sun.[52]

He seems, then, to be using Christianity in a symbolic sense, so that 'the way of nature is, in this deep, though earthen sense, the way of the Cross'.[53]

I mentioned earlier the possible connections between Holmes Rolston III's approach and Gaian spirituality. The Gaian image, which also carries with it connotations of the goddess, automatically brings a theological dimension into the discussion of environmental ethics. Lovelock has identified Gaia with a Marian spirituality, finding in Mary a symbol for Gaia. While he argued at great length that his hypothesis was primarily scientific, rather than religious, the spiritual implications of his idea still surfaced in his writing. He asks us to consider:

> What if Mary is another name for *Gaia*? She is of this Universe and conceivably, a part of God. On Earth she is the source of life everlasting and is alive now; she gave birth to humankind and we are a part of her . . . That is why, for me, *Gaia* is a religious as well as a scientific concept . . .[54]

It seems to me that Lovelock is using Mary in a symbolic sense to express his own Gaian spirituality, just as Rolston uses the cross to express his ideas. They both seem to be pointing to an originate form of spirituality which they believe surfaces, even if at times in a hazy way, in folk religion.

Anthony Weston has commented on Gaia as 'extraordinarily suggestive for environmental philosophy'.[55] Stephen Clark considers the impact of Gaia on our lifestyle.[56] He is cautious in assuming the earth is an 'organism', but suggests that there are heuristic advantages in seeing the earth as an entity.[57] I would agree with Clark thus far. He also suggests that there is no need to consider that there is a 'vital centre' of the earth; in other words that Gaia has some kind of consciousness. While this is perfectly true from a scientific perspective, the fact remains that Lovelock assumes that there is a directedness or teleology in Gaia, that is the survival of life. While he has provided evidence that this process is a result of automatic feedback, from his 'daisyworld' model, I have argued elsewhere that such a model does not 'prove' the process is automatic.[58] I could equally devise an automated biological model which is destabilizing. The question as to why the directedness of Gaia is towards the maintenance of life is unanswered. His hypothesis implies a form of consciousness, even though he does not spell this out. Moreover, it is far from clear what this life might be, if it is the life of simple bacteria and micro-organisms, which Lovelock's theory might suggest, the idea of Gaia as a 'model of cooperation' starts to become a trifle ludicrous.

Whatever the outcomes of the scientific debates, it seems to me that one of the attractions of the Gaia hypothesis is the seeming affirmation of a spiritual dimension within the world process. Anne Primavesi and Rosemary Radford Ruether have used the idea of Gaia to stress the feminine divine in nature. Ruether, for example, describes the divine in Gaia as the 'true source of life' and 'that power from which the universe arises', speaking from the 'intimate heart of matter'.[59] For her it is primarily the perception of God in male terms alongside a domination of the earth that has contributed to the ecological crisis. Our lifestyle, and by implication ethics, demands a reinterpretation of the notion of covenant and sacrament. For her covenant expresses the relationship of all living things to God, while sacrament links human to all other life forms in a 'dance of energy'. Anne Primavesi, too, adopts an anti-androcentric view, rejects 'dualist' ideas and correlates ecology with patterns of relationships. Her theology is also one which tends towards mergence. Hence: 'This experience of the Spirit springs from one single reality: God and creation united . . . There is no separation between our love for one another and our concern for the well-being of the ecological community of human and non-human species.'[60]

It is this seeming total collapse of distinctions between God and the world, human and non-human, self and other, found alike in deep ecology, much eco-feminist writing and neo-Gaian philosophy that I find disturbing. Process theology, too, tends in the same direction. However, in this case the universalizing tendency is towards greater 'enrichment', though it is unclear what this 'enrichment' might entail. I suspect that it is of a particular anthropocentric variety. Process theology reacts against the so-called 'scholastic' model of God, seen as remote, timeless and unchanging. For process theology the world becomes God's body, or perhaps God in his corporal being. Process theology differs from pantheism in that any change is said to originate in God.[61] Yet it seems to me that once the identification of God and the world becomes too strong, the whole notion of what it means to be God becomes fraught with difficulty. While I would share process theology's notion of a God who, in some sense, suffers with us, the idea that it is identical to our suffering seems too crude. As Clark has pointed out:

> If there is no eternal truth or standard, then God's changes, consequent upon the changes of his elements are just that, mere changes. The doctrine is proposed to allow for real progress in the world, but instead we are bound to conclude that what seems to us corruption is as likely as 'moral development', and that there is no lasting criterion anyway by which corruption and progress are distinguishable. The suffering with us of such a being is no consolation.[62]

I would also tend to agree with Clark that process theology, by claiming that the teleology of the earth is towards the divine Eros, fosters the myth of progress. The dream of hominization, spelt out in the theology of Teilhard de Chardin, is not far beneath the surface. Hence, far from encouraging a greater appreciation of the natural world, such panpsychism leads to a subtle form of anthropocentrism.

The way forward, it seems to me, is to regain a sense of the Other in God and between God and the natural world. This means treating the world with respect, not as an 'It', but as an 'Other', the 'I' and 'Thou' of Martin

Buber.[63] I also believe that an understanding of a remote, domineering God of power is no longer tenable. Yet, a stress on the immanence of God in creation does not necessarily mean that we can have no concept of transcendence. Similarly, the realization that we are like animals in many respects need not obliterate any distinctiveness.

Gustafson has pointed out that religion can at times be propagated for the purposes it serves in sustaining moral causes. Whatever brings humans happiness is assumed to be right, so that God becomes the instrument of humanity.[64] He poses the dilemma: Do we take as a starting point the critical social issues of our time, so that these frame the questions? Or do we take as a starting point the knowledge of God and let religious beliefs and outlooks shape the questions and answers? I suggest that an exclusive reliance on either approach is unjustified. Rather the two methods can come together in fruitful exchange. Environmental ethics is one area where the critical social issues do ask specific questions which do not stem readily from theology alone. On the other hand, I have tried to show in this chapter that an internal discussion remains incomplete. Theology does have something to contribute to the debate, but it is there to challenge and foster dialogue, rather than to support preconceived ideas. The language of theology, unlike the language of ecology, remains metaphorical. While deep ecology has sought for meaning, theology is at home in the world of searching after meaning. It is true that if we opt for specific solutions to specific ethical questions through casuistry, more general issues of belief can seem irrelevant. However, I intend in the following chapter to show how particular issues can raise questions which are important springboards for theological discussion.

For the time being it is the challenge to the overall ethos of modern culture which strikes me as an important positive contribution of theology to environmental ethics. The solution of environmental philosophers seems to take two different paths. Either a pragmatic goal is sought or there are plural solutions to plural questions. The absence of any real criteria becomes obvious. The metaphilosophical option, which still allows in theory for foundations, does not seem to me to make any real practical difference. The alternative to this pragmatic goal is the 'pie in the sky' mysticism of deep ecology, a mergence of the self with all that is, a cosmic interrelatedness of us to all of life, a fusion with the wild in natural systems. But now what effect does this have other than either encouraging us to ignore evil altogether or to suffer further through self-identification with the suffering of the planet? Identity may increase our sensitivity to the environmental burden, but it certainly does not offer practical reasons for hope. Indeed, it seems to instill a false hope that our own enrichment is equivalent to the enrichment of the earth, thus bringing back the feared anthropocentrism that is so vigorously attacked. This anthropocentrism seems too deeply embedded, set in motion by post-Enlightenment culture and thereby making humanity 'the measure of all things'.[65]

Historically, moral theologians have assumed that what is good for humanity must be good for creation; that the world is created with humankind in view. It seems to me that this idea subsequently took on a secular form through the Anthropic principle.[66] Yet to assume that the orientation of God's purpose is exclusively towards human well-being is, it seems to me, to take a view which makes God in our own image. Rather, the idea that humanity is made in the image of God suggests shared responsibility of humans for all creation, rather than a total identification. Nonetheless, I would agree with Francis Watson that interpretation of the image of God needs to go beyond simple functional categories.[67] One strand of this idea is that of praise. Humanity is unique in the praise offered to God on behalf of creation. Yet the New Testament takes us beyond this idea, though it does not to my mind exclude it, which Watson suggests. Christ is the transformed image of God, which shows humanity what it means to be that image.

Watson suggests that the use of Psalm 8 by the letter to the Hebrews, and by implication Genesis 1:26–8, is reinterpreted so that Christ is now Lord of all, rather than humanity alone as lord of creation. In other words, our dominion of the earth is replaced by the lordship of Christ, expressed not as domination, but in loving humility. I find this idea fascinating. The inspiration for theocentric ethics has tended to draw exclusively on the Genesis account, pointing out that our dominion of the earth is really best seen in terms of stewardship. We become the caretakers of God's creation. But in many respects this idea of stewardship is not really adequate.[68] Furthermore, the idea of the cosmic Christ could suggest a Christ figure who is remote from our experience. If we identify the image of God with Christ through use of the Hebrews passage, rather than Colossians, this seems to have two consequences. Firstly, all of creation is in subjection to Christ, but it is one based on love and humility rather than power and oppression. Secondly, Christ becomes the model for our action in the world. We 'learn in him what is the true nature of the rule over the created order that was promised in the beginning'.[69]

The consequences for environmental ethics means that our treatment of nature comes from an attitude of humility, rather than the stewardship ideal which still tends towards 'resource management'. In practice this means a lack of interference as much as creative involvement. It seems to me that the means through which this becomes possible cannot rest in Christology alone. Rather, the Holy Spirit's work in us is the means for effective transformation to the true image of God. This same Spirit is also at work in all of creation, so I would wish to keep the idea of all of creation in some sense giving praise to God; human praise can merely echo the praises of all of the natural world. I am not suggesting thereby that all of creation is somehow 'humanized'. I think this is possibly the danger of a cosmic Christology which moves out to the world from the cross. Rather, all of creation offers praise to God as Other than human, but sharing in some sense in the same Spirit, who also identifies with the divine Spirit.

I prefer then, a trinitarian expression of the relation between God and creation, which has implications for environmental ethics. The Spirit at work in the world can remind us of our interconnectedness with everything else. But the figure of Christ shows the divine humility that is to become ours if we are to truly reflect the divine image. Our treatment of the earth must, it seems to me, stem from an attitude of repentance and humility. It is respect, rather than reverence for the natural world. While reverence leads to passivity, respect 'implies a receptivity on our part, a taking in, an attentiveness and openness with regard to the world around us'.[70] Cowdin resists the idea of 'radical self-sacrifice' in our relationship with the natural world as 'inappropriate behaviour'. I am less convinced. Unless we aim at least towards a radical self-sacrifice, it is all too easy to rationalize our own abuses. The cross is the divine Wisdom of God that confounds human wisdom. But the cross is not all there is to say. The Wisdom of God includes in the cross the hope of the resurrection, the hope of shared praise of all of creation.

It is suffice to say that the eschatological nature of the Christ event reminds us of both the cost to creation and the future promise of heaven. What kind of heaven can we envisage? Do animals have souls and, if so, should this alter our treatment of them? The affirmation in Genesis of all of creation as good has led many to question the belief that animals are made for human benefit. As Margaret Atkins suggests: 'Is it not reasonable to hope that the final consummation will be one that preserves all that is good, redeems rather than annihilates God's flawed creatures.'[71] Others have argued that the mortality of plants and animals is an inevitable part of their existence.[72] Yet who is to say what is inevitable in the future life? I would tend to resist any ideas of a radical discontinuity. The story of creation in all its manifold richness and variability, beauty and awe may have some future existence. If we refuse to admit this then all our striving in environmental ethics starts to seem pointless. If the future is just one for humans alone, then there seems little incentive in preserving and fostering all life forms on earth.

I would agree with Atkins that it is hard for us to imagine what immortal life is like for animals. But then I cannot imagine that they or us for that matter would be satisfied in a barren landscape, with the humbler life forms ignored and forgotten. While I would resist spelling out a portrait of heaven that gives a picture of a future life in the 'new cosmos', I believe that our hope and the hope of all creation is bound up together. This does not remove the distinctiveness of humanity. On the contrary, our distinctiveness and theirs serve to make a heaven that would be far less boring than one full simply of rational creatures. It is not just the rational self that responds to God, but the whole self, mind, body and emotions. Do animals have emotions? A growing body of literature now answers in the affirmative.[73] Yet for me it does not really matter whether non-human life has emotions or not; the point is that each in its own way will find repose in God. In some sense the natural order has to undergo transformation, as the vision of Isaiah suggests.

For the time being we can think of all life forms as companions, sharing in some sense our biological existence, but remaining radically different from us. Cowdin resists any notion of extending a covenant to nature on the basis that 'human categories are misused'.[74] Yet it seems to me that there is a scriptural basis for a covenant with the natural world in the Noahic covenant of God with all creation. Humans, in the image of God, share in this covenant. The idea of covenant does not mean that all covenants are identical. Or that the two parties are equivalent. The covenant between God and humanity makes this clear enough. I share Ruth Page's idea of the companionship of animals, though I would wish to extend this to all creation. As she points out, in this present world this leaves us with some tough choices: 'Some freedoms are diminished, as in the case of farm animals, and some possibilities extinguished as in the case of the diphtheria virus, while we hope that the HIV virus will follow its extinction.'[75] This is qualified by the fact that, as Christians, we cannot 'bracket off some of creation, call it "vermin" and cease to care'.[76]

Killing vivisection scientists is certainly not justified on the basis of their cruelty to animals. A Christian view seeks for a change of heart, a *metanoia*, which applies to any human being. It seems to me that there can be no 'just war' in favour of animals, at least not one which threatens human life. The message of Christ is the message of peaceful negotiation. The equivalence of human and animal life envisioned by some environmental philosophers, by implication at least, would seem to condone rather than resist such practices.

Stephen Clark makes an important point when he states that Nature is our sister, rather than our mother; an object of respect, not reverence.[77] Are there just two alternatives: submission to Nature or submission to humanity? But even the former is muddied as, if we just do what comes naturally, we end up following any and every instinct to rampage the earth. The challenge of an eschatological approach to our behaviour now is very real. As Clark suggests: 'It might nonetheless be wise to ask ourselves how we shall greet God's creatures in the Promised Kingdom: the creatures, that is, whom we have paid to deform, degrade, deprive, depress, destroy, for the sake of transitory pleasures in the here-and-now.'[78]

Conclusions

I have tried to show the spectrum of interpretations of value in the natural world according to environmental philosophy. While the move away from anthropocentrism is welcome, where this implies an ethic for human benefit alone, the reverse tendency towards an ecofascism is to be resisted. This fascism takes shape either in the extreme responses of animal rights activists who seek the killing of the humans responsible, or in the oppressive image of the Earth Mother as vengeful goddess, waiting to remove the cancerous growth of humanity at large. Instead a theocentric ethic, rather than being shaped by the world of nature and the evolutionary myth of progress, as in process theology, needs to take its bearings from affirmation of the goodness of creation and the recognition of God and the world as Other. This otherness implies a 'Thou', rather than an 'It', so that respect for the natural world flows from a recognition of the earth as gift.

The future of all creation is one which we can look forward to in hope, but it is a hope that includes all of creation, not just rational beings. This hope challenges us to work for transformation now, but it is a transformation more of our own attitudes towards humility and respect, rather than an egocentric domination of the earth. How can we seek to reconcile justice in the human community with justice for all creation? How does the reality of environmental ethics, which is bound up with issues of poverty and deprivation, temper our understanding of how to treat the earth? In other words what is the relationship between developmental and environmental ethics? In the next two chapters I will begin to address these questions from within a theological perspective.

NOTES

1. E. C. Hargrove, *Foundations of Environmental Ethics* (Prentice Hall, Englewood Cliffs, 1989); R. Attfield, 'Has the History of Philosophy Ruined the Environment?' in *Environmental Philosophy: Principles and Prospects*, R. Attfield (ed.) (Avebury, Aldershot, 1994), pp. 77–87.

2. A. Light and E. Katz, 'Introduction: Environmental Pragmatism and Environmental Ethics as Contested Terrain', in *Environmental Pragmatism*, A. Light and E. Katz (eds) (Routledge, London, 1996), p. 1.
3. J. Passmore, *Man's Responsibility for Nature*, 2nd edn (Duckworth, London, 1980).
4. See, for example, P. Singer, *Animal Liberation: A New Ethics for Our Treatment of Animals* (Jonathan Cape, London, 1986); T. Regan and P. Singer (eds), *Animal Rights and Human Obligations* (Prentice Hall, Englewood Cliffs, 1976); T. Regan, *The Case for Animal Rights* (University of California Press, Berkeley, 1983).
5. A. Schweitzer, *Civilisation and Ethics* (1923), trans. C. T. Campion (Unwin Books, London, 1967 edition).
6. P. Taylor, *Respect for Nature: A Theory of Environmental Ethics* (Princetown University Press, Princetown, New Jersey, 1986).
7. *Ibid.*, p. 72.
8. *Ibid.*, p. 90.
9. *Ibid.*, p. 115.
10. *Ibid.* See the section on 'Competing claims and priority principles', pp. 256ff.
11. H. Rolston III, *Environmental Ethics: Duties to and Values in the Natural World* (Temple University Press, Philadelphia, 1988), p. 172.
12. *Ibid.*, p. 188.
13. H. Rolston III, 'Values in Nature and the Nature of Value', in *Philosophy and the Natural Environment*, R. Attfield and A. Belsey (eds) (Cambridge University Press, Cambridge, 1994), p. 19.
14. *Ibid.*, p. 26.
15. *Ibid.*, p. 28.
16. J. Lovelock, *Gaia* (Oxford University Press, Oxford, 1979); J. Lovelock, *The Ages of Gaia* (Oxford University Press, Oxford, 1988); C. Deane-Drummond, 'God and *Gaia*: Myth or Reality?', *Theology* (July/August 1992), pp. 277–85.
17. K. Pedler, *The Quest for Gaia* (Harper Collins, London, 1991). For further development and discussion of ethics and Gaia, see C. Deane-Drummond, 'Gaia as Science Made Myth: Implications for Environmental Ethics', *Studies in Christian Ethics* (Summer 1996), pp. 1–15.
18. A. Leopold, *A Sand County Almanac* (Oxford University Press, Oxford, 1949).
19. J. B. Callicott, *In Defence of the Land Ethic* (State University of New York Press, New York, 1989).
20. E. Sahouris, *Gaia: The Human Journey from Chaos to Cosmos* (Simon and Schuster, New York, 1989).
21. See, for example, K. Pedler, *op. cit.*, pp. 9–10.
22. Warwick Fox argues that an altered consciousness is all that is required and as such makes ethics superfluous. W. Fox, *Towards a Transpersonal Ecology* (Shambala, New York, 1991), p. 225.
23. K. Pedler, *op. cit.*, pp. 173–4.
24. J. Milbank, 'Out of the Greenhouse', *New Blackfriars* (January 1993), pp. 4–14.
25. J. Lovelock, 'Planetary Medicine: Stewards or Partners on Earth?', *The Times Literary Supplement* (13 September 1991), pp. 7–8.
26. S. Clark, 'Global Religion' in R. Attfield and A. Belsey (eds), *op. cit.*, p. 114.
27. P. Singer, *op. cit.*
28. T. Regan, *op. cit.*
29. T. Regan, 'The Case for Animal Rights', in *People, Penguins and Plastic Trees: Basic Issues in Environmental Ethics*, D. van de Veer and C. Pierce (eds) (Wadsworth, Belmont, USA, 1986), p. 39.
30. D. van de Veer, 'Interspecific Justice' in D. van de Veer and C. Pierce (eds), *ibid.*, pp. 51–6.
31. *Ibid.*, p. 56.
32. L. Johnson, *A Morally Deep World: An Essay on Moral Significance and Environmental Ethics* (Cambridge University Press, Cambridge, 1991).
33. *Ibid.*, p. 178.
34. S. Clark, *The Moral Status of Animals* (Oxford University Press, Oxford, 1984), p. 120.
35. A. Linzey, *Animal Theology* (SCM Press, London, 1994). For additional discussion of the use of 'rights' language, see M. Midgley, 'Animals and Why They Matter', *Theology in Green*, vol. 5, issue 2 (Summer 1995), pp. 22–32.

36. E. Graham, *Making the Difference: Gender, Personhood and Theology* (Mowbray, London, 1995).
37. S. Clark, 'Global Religion', in *op. cit.*, p. 126.
38. M. Midgley, 'The End of Anthropocentrism', in R. Attfield and A. Belsey (eds), *op. cit.*, pp. 103–12.
39. *Ibid.*, p. 111.
40. *Ibid.*, p. 112.
41. See, for example, R. J. Bernstein, *The New Constellation* (Polity Press, Cambridge, 1991).
42. See especially K. A. Parker, 'Pragmatism and Environmental Thought', in A. Light and E. Katz (eds), *op. cit.*, pp. 21–37.
43. A. Light and E. Katz, 'Introduction: Environmental Pragmatism and Environmental Ethics as Contested Domain', in A. Light and E. Katz (eds), *op. cit.*, pp. 1–18.
44. A. Weston, 'Beyond Intrinsic Value: Pragmatism in Environmental Ethics', in A. Light and E. Katz (eds), *op. cit.*, pp. 285–306.
45. J. Baird Callicott, 'The Case Against Moral Pluralism', *Environmental Ethics*, vol. 12 (Summer 1990), pp. 99–124.
46. K. A. Parker, *op. cit.*, esp. pp. 31–5.
47. A. Light, 'Environmental Pragmatism as Philosophy or Metaphilosophy? On the Weston-Katz Debate', in A. Light and E. Katz (eds), *op. cit.*, pp. 325–38.
48. *Ibid.*, p. 330.
49. W. K. Frankena, 'Ethics and the Environment', in *Ethics and the Problems of the 21st Century*, K. E. Goodpaster and K. M. Sayre (eds) (University of Notre Dame Press, Notre Dame, 1979), pp. 3–20.
50. H. Rolston III, *Philosophy Gone Wild: Environmental Ethics* (Prometheus Books, Buffalo, 1989).
51. *Ibid.*, p. 221.
52. *Ibid.*, p. 257.
53. *Ibid.*, p. 259.
54. J. Lovelock, *The Ages of Gaia*, p. 206.
55. A. Weston, 'Forms of Gaian Ethics', *Environmental Ethics*, vol. 9 (3) (1987), p. 217.
56. S. Clark, 'Gaia and the Forms of Life' in *Environmental Philosophy: A Collection of Readings*, R. Elliot and A. Gare (eds) (Open University Press, Milton Keynes, 1983), pp. 183–94.
57. S. Clark, *How to Think About the Earth* (Mowbray, London, 1993), p. 26.
58. C. Deane-Drummond, 'Gaia as Science Made Myth'.
59. R. Radford Ruether, *God and Gaia: An Ecofeminist Theology of Earth Healing* (SCM Press, London, 1993), pp. 5, 227, 254.
60. A. Primavesi, *From Apocalypse to Genesis: Ecology, Feminism and Christianity* (Burns and Oates, Tunbridge Wells, 1991), pp. 263–4.
61. See S. Clark, *How to Think About the Earth*, p. 78. The idea of Charles Hartshorne that God has an eternal aspect seems to counter any charge of process theology as pantheistic. However, I think Clark is correct to suspect that this subtlety is ignored by many of those who claim to identify with process thought, see p. 84.
62. *Ibid.*, p. 83.
63. See A. Hodes, *Encounter with Martin Buber* (Penguin Books, London, 1975).
64. J. M. Gustafson, *Theology and Ethics* (Blackwell, Oxford, 1981), pp. 19–20.
65. *Ibid.*, p. 81.
66. J. D. Barrow and F. J. Tipler, *The Anthropic Cosmological Principle* (Oxford University Press, Oxford, 1986).
67. F. Watson, 'In the Image of God', paper delivered at *The Society for the Study of Theology* (April 1996), pp. 1–23.
68. For a further discussion of some of the reasons for this inadequacy, see C. Palmer, 'Stewardship: A Case Study in Environmental Ethics' in *The Earth Beneath: A Critical Guide to Green Theology*, I. Ball, M. Goodall, C. Palmer and J. Reader (eds) (SPCK, London, 1992), pp. 67–86.
69. F. Watson, *op. cit.*, p. 17.
70. D. M. Cowdin, 'Toward an Environmental Ethic', in *Preserving the Creation: Environmental Theology and Ethics*, K. W. Irwin and E. D. Pellegrino (eds) (Georgetown University Press, Washington, 1994), p. 135.

71. M. Atkins, 'Could There be Squirrels in Heaven?', *Theology in Green*, issue 4 (October 1992), p. 21.
72. *Ibid.*, p. 21.
73. See M. Midgley, 'Animals and Why They Matter', *Theology in Green*, vol. 5, issue 2 (Summer 1995), pp. 22–32.
74. D. M. Cowdin, *op. cit.*, pp. 140–1. As one might expect he also rejects the idea of extending rights language to animals for much the same reasons.
75. R. Page, 'The Animal Kingdom and the Kingdom of God' in Occasional Paper No. 26, *The Animal Kingdom and the Kingdom of God* (The Church and Nation Committee of the Church of Scotland and the Centre for Theology and Public Issues, Edinburgh, 1991), p. 7.
76. *Ibid.*, p. 8.
77. S. Clark, 'Ecology and the Transformation of Nature', *Theology in Green* (October 1995), p. 33.
78. *Ibid.*, p. 45.

4

Biotechnology: an ethical critique[1]

The ethics of genetic engineering, or 'genethics' in current jargon, is one of those areas of debate which needs to be constantly reviewed in the light of the rapidly expanding science of biotechnology. While biotechnology is the general application of biology to commercial practice, genetic engineering is one branch of this practice. The revolution taking place within biology may be as significant for our existence as Newton's and Einstein's ideas within physics. The possibility of manipulation of human genes has led to considerable discussion and debate amongst moral philosophers and theologians.[2] The term biotechnology usually implies genetic engineering when used in ethical discussion. I intend to examine the relatively neglected area of genetic engineering as used for agricultural purposes. The philosophical, theological and ethical implications of this application of biology are rather different from that pertaining to human beings and deserve separate attention.

In particular, this technology has important environmental consequences, both in the long-term and the short-term. The possible long-term effects on the human community are equally significant, both directly in terms of North–South relationships and indirectly through environmental influences. It seems to me that these potential effects need to be considered a priori if environmental ethics is to have any real bearing on future policy decisions. I intend to draw on the theoretical framework of the last chapter as well as show how the issues raised by agrobiotechnology both challenge and are challenged by this framework. I hope to show that such an approach combines some of the advantages of environmental pragmatism, in its aim to be rooted in experience, while drawing on a theological framework for ethics. This framework is one which is sufficiently flexible to take account of this experience, but is bold in its critique of unethical practice.

The tendency amongst theologians and philosophers is to adopt what William Temple described as a 'middle axiom' approach when reflecting theologically on social issues.[3] This is the establishment of broad theological principles, while at the same time resisting any attempt to make more detailed recommendations. In some ways this seems to be a valid approach since there are relatively few theologians and philosophers who are equipped to understand the detailed knowledge of science that is needed in order to make a realistic contribution to scientific practice. Furthermore, experts tend to resent their territory being 'invaded' by outsiders. It is all well and good to make general comments, which can then be safely ignored, but specific proposals and suggestions are considered to be out of bounds. This, in effect, inevitably makes the philosophical and theological writing sound good to the public ear, but irrelevant to actual practice. I am not suggesting that there is no place for such overviews. Stephen Clark's book, *How to Think About the Earth*, for example, more than achieves his modest aim to be a serious critique of the larger claims made by preachers, lobbyists and politicians.[4] Ian Barbour, on the other hand, hopes to give his readers fundamental insights into biology by giving an outline of the basic science of genetic engineering in his book, *Ethics in an Age of Technology*.[5] However, the philosophical and theological implications are related to technology in general, rather than the specific issues associated with genetic engineering.

My purpose here is to look at three different attitudes to genetic engineering that are most common, using particular case studies as illustrations. I will then offer a philosophical critique in the light of real decisions faced by both policy makers and scientists. Finally, I will indicate ways in which theological reflection can contribute to ongoing discussion. My method, then, is 'from below', drawing on particular examples which are environmentally significant as a basis for reflection. This mirrors, in part, my own experience of biologist first, followed by theologian. As I indicated in the last chapter my approach is based on certain theological premises, such as belief in the goodness of God, the love of God for all creation and the idea of humankind made in God's image.[6]

Genetic engineering as promise

The green revolution is the cultural and scientific background which spawned an optimistic attitude to genetic engineering as applied to agriculture. In the late 1960s traditional plant breeding methods were used by scientists to develop high-yield varieties. When this was combined with more intensive use of fertilizer, there were vast increases in production, especially in Third World countries. In India, for example, the wheat crop was doubled in six years.[7] However, the dream that the green revolution could solve the world food crisis came up with some unexpected difficulties. Small farmers and the rural communities were ousted by wealthier landowners who could afford the costly fertilizers. In addition, the mechanized form of farming reduced the need for human labour and thus, ironically, increased overall poverty and deprivation. It is ironical that while India has now become a food exporter, the poor still cannot afford the food they need for survival. There are, of course many other factors which contribute to malnutrition in the Third World, such as an over-concentration on export crops, rather than staple food crops. In this case the balance of trade favours the consumers.[8] A detailed treatment of trade deficits and other general factors, such as overpopulation, which indirectly lead to malnutrition is outside the scope of this book. However, I will be returning to a general discussion of these issues in the next chapter.

More recently there has been a drive for the development of crops which require fewer fertilizers as well as the introduction of appropriate technology that requires less fuel.[9] Much of the task of genetic engineering has been no more than to complement that of traditional breeding methods. However, instead of taking several years to develop a new crop, it now takes a matter of months. In this respect genetic engineering could be seen as a liberation from time constraints imposed by slow-growing crops. In a hungry world few would wish to legislate against the development of crops that could flourish in the poorer areas of the world hampered by dry, salty or nutritionally poor conditions. However, while the green revolution was directed towards the needs of the Third World, at least in its original intention, contemporary genetic engineering is more often conducted in the West under industrial contracts requiring expensive patents for new varieties. The perceived novelty of genetic engineering techniques and the organisms they produce has allowed researchers to patent living organisms, their parts or processes. The first patent was given in 1980 by the Supreme Court in the USA for a genetically modified bacterium. Patenting has accelerated the trend towards monopolies in seed production by one or two large companies which, according to Barbour, 'reduces both genetic diversity and economic competition'.[10]

In addition to this disturbing loss of genetic diversity, the research aims at maximum profits. These are to be found in projects such as the genetic engineering of bacteria to synthesize prescription drugs or the modification of crops for the Western market to maximize profit. We only hear about a small fraction of these developments in the media. The recent scenario of a genetically engineered tomato, designed so that it is said to have an improved flavour and now available in processed form in supermarkets, is just the tip of the iceberg. Another genetically engineered tomato known as FLAVR SAVR may soon be on sale in the UK.[11] This tomato was launched amidst a wave of publicity in the USA as the first genetically engineered whole food. This tomato has the gene which leads to softening 'scrambled', so that the tomato ripens without going squashy. The engineered tomato allowed companies to transport tomatoes long distances following ripening, instead of artificially ripening green tomatoes. This was to the commercial advantage of the producers. The research was directed not so much for the consumers' benefit as for the commercial gain of producers. More to the point, perhaps, the worry that the tomato might not be 'safe' is largely illusory in this particular case. What was missed is the concentration of money and research on development of luxury items for the Western market. This, to my mind at least, is more insidious as it is seemingly hidden. The promise of genetic engineering seems, then, to be directed towards the needs of high-income populations. This leads into a discussion of the way this technology can be a weapon of power, which I turn to next.

Genetic engineering as power

A specific example of the distribution of research to date is that more money has been spent on the development of strawberries that can withstand frost conditions for the spring USA market than on improving the yield of basic subsistence crops, such as cassava, maize or bean plants in the Third World.[12] Other rapidly expanding technologies include the development of tissue culture grown in laboratory conditions that have been engineered to produce 'synthetic' products. It may be a matter of time before a biotechnological means is found to produce substitutes for substances such as vanilla or cocoa. If this were to take place we would witness a collapse in the economy of Madagascar, which relies on vanilla bean exports, and in the economy of West Africa, which relies on cocoa. Biotechnology is becoming a means of oppressing Third World economies and seems to drive a deepening wedge between rich and poor nations. It is these long-term *social* consequences of genetic engineering which need careful consideration. While there is legislation in place, at least in principle, to protect us from possible health risks of genetic engineering, it becomes much harder to legislate towards research priorities. The market economy seems too crude an instrument to act in a way which protects the interests of all those concerned. The promise of genetic engineering is becoming, instead, a means for making a profit in a way that allows further domination of poorer Southern nations by the richer Northern ones.

There is another sense in which genetic engineering can be seen as a power, that is human power and domination of the natural world. It is now possible, for example, to genetically engineer bovine growth hormone (BST) in laboratory conditions.[13] If this is injected into cows, their milk production increases by 15 to 20 per cent. There is considerable controversy over the exact nature of BST, that is whether it is identical to the hormone produced naturally. There are also arguments about the potential stressful effects on cows, such as an increase in incidence of mastisis, which has attracted considerable attention from those concerned with animal welfare. Moreover, the use of BST is confined to large farms, which puts at a disadvantage farmers with small acreage using traditional methods. It is also a potential risk to human health as its long-term effects have not been properly investigated. Overall, it is ironical and unnecessary, given the overproduction of milk and excess of dairy products in the Western market. The idea that cows and other farm animals can be manipulated in this way for human benefit alone, encourages the human perception of animals as resources to be managed; animals become simply 'biomachines'.

Even more insidious developments include research into 'production' of animals that can withstand overcrowded conditions.[14] Although conventional breeding in this case is used to achieve this aim, it may just be a matter of time before the same result could follow from genetic engineering. In this case it represents a loss of a particular capacity, namely a response to crowding, which would be easier to achieve by genetic engineering compared with an addition of an attribute. This is related to the fact that deletion of sections of genetic material is easier than unravelling the regulation of a complex battery of genes that are required for certain character attributes. In most cases, however, deletion is likely to be lethal and thus unusable. However, even if side-effects arose, such as loss of intelligence in pigs, such effects would be ignored as long as the production of lean meat was unaffected. The attempt to channel research in this direction exemplifies the bland assumption that the animals are little more than a mechanism to be manipulated for human profit. The aim seems to be to produce an animal that is merely a passive 'vegetable' in intensive conditions, which amounts to a cruel loss in quality of life. This same attitude has led to an increase in the much publicized 'mad cow disease', with its controversial link with the human equivalent brain disease. Whether the disease in cows is 'caused' by feedstuffs containing animal products, or some other general component, such as organophosphates, does not seem to me to be really relevant. The point is that the industry, in aiming at maximum profit, has no sense of the intrinsic value of these creatures. Indeed, it seems to me that it is only by treating the animals as a resource to be managed that factory farming can be carried out with a clear conscience.

Genetic engineering as threat

The genetic engineering of animals in the manner described above could be seen as a threat to animal welfare for the sake of human interests. However, the fear of genetic engineering in the public mind is more often associated with a perception of either the health risk of genetic engineering *per se*, or the threat to authentic human existence.

The environmental risk factors associated with genetic engineering are related to the power of the technology to bring about irreversible change in the hereditary material of plants, animals and bacteria. For example, it is now possible to genetically engineer crop plants that are resistant to chemical herbicides. A government advisory committee has given a Belgian company permission to release into the market a genetically engineered rapeseed that contains resistance to the herbicide 'Basta'.[15] This allows the farmers to control weeds in rapeseed fields that would naturally be susceptible to the herbicide. There were no public consultations prior to the decision by the committee. In this case the risk factors will be enhanced because the environmental consequences are not monitored adequately by a company determined to profit from the product. While the company denies any risk to the environment, it is not proven that this rapeseed will be benign in an environmental sense. If anything, the science suggests the opposite, since rapeseed can cross-fertilize with wild mustard plants and even become a weed on roadside verges. Once these genetically engineered plants became established they would be difficult to control as they would be herbicide resistant. It is a sobering fact that in Britain there have been over sixty small-scale releases of genetically engineered plants, but there has been only one instance of adequate scientific study of the environmental consequences.

For the above example there are other, more indirect risk factors. The very development of these varieties encourages heavy use of herbicides, which are themselves a threat to natural ecosystems. The long-term effects on human health are unknown. Government regulations rarely consider indirect risks of this type. Furthermore, the increased dependence of the farmers on herbicides for weed control encourages an equal dependence on the hybrid seed sold by the same company as one package. Hybrid seeds do not breed true to the next generation, which means that the next generation of plants is not uniform. Hybrid plants are the result of a cross between two related varieties, so that the seeds in the next generation are mixed genetically. This variation leads to a drop in yield for a variety of reasons. For example, farmers rely on mechanized means of harvesting, which in turn makes them dependent on a crop that is uniform in height and time of ripening in order to achieve high yields. The traditional method of saving seed for the next year's crop is now impossible, or rather only possible with loss of yield. The farmers are thus forced to buy new seed and herbicide every year.[16] As David King has recently pointed out:

> Standardized farming means that companies can pursue global marketing strategies based on a few different seed varieties and herbicides, and reap the benefits of the massive economies of scale which have always been one source of their competitive advantage. Herbicide tolerance is a prime example of the way that industrial companies are pursuing their long term aim of profitability, by moulding agriculture to their own image. The significance of genetic engineering is that it gives the companies powers that were never before available, to adapt what the farmer sows to their, rather than the farmer's needs.[17]

Another indirect risk, which is also characteristic of conventional plant breeding, is the overall loss in genetic diversity, to which I alluded briefly above. However, in the case of genetically engineered plants this uniformity can be transferred to other species as well through a technique known as 'transgenic manipulation'. A Dutch company has tested the herbicide resistance for rapeseed, discussed above, for use in chicory plants. In this case a section of genetic material which contains the herbicide resistance is introduced into the chicory plants. Crops which grow from wild strains have a much greater variability which protects them naturally from pests and disease. When a crop is genetically engineered the resultant uniformity brings the desired increase in yield but also carries a greater vulnerability to disease. Much of the classical plant breeding was directed towards keeping 'one step ahead' of the rapidly changing populations of virulent fungi and other pests. While the genetic engineering of plants can speed up this search for resistant crops, it eventually leads to a loss of variability needed for potential change. This loss of variability within one species is irreversible. In order to find new sources of variation researchers have sought wild strains which have retained their genetic variability. These wild strains are on the whole confined to the poorer Southern continents. The need for foreign investments means that in many cases the patent charged for these seeds do not reflect the potential benefit for Northern markets.[18] The threat of genetic engineering becomes, in this case, another occasion for the abuse of power.

Another risk of transgenic manipulation which is often not publicly acknowledged is that new proteins in the transgenic food may cause potentially lethal reactions in humans. For example, those people who are allergic to peanuts may be careful to avoid eating peanuts or products containing nuts, but if they eat an unlabelled genetically engineered plant of another species which contains nut proteins, they may suffer fatal allergic reactions. The US seed company Pioneer Hi Bred, for example, was forced to drop development of genetically engineered soybeans containing Brazil nut genes. Extracts of the genetically engineered soybeans reacted with human blood serum from individuals who suffered from nut allergy.[19] While it might be possible to exclude certain plants which are known to cause allergic reactions, many foods are allergenic to a smaller number of people. In addition, new proteins from bacteria and insects which are not normally consumed by humans could lead to hitherto unknown allergenic reactions. Other, more subtle effects on the function of the genetic material of the transgenic organism could lead to the production of new toxins which are not present in the 'pure' strain.

The transgenic manipulation of animals raises further issues of animal welfare. For example, pigs have been injected with human growth hormone genes, and sheep with genes from bacteria. How many genes can we introduce before it becomes a new species? Is it right to violate the 'pigness' of a pig?[20] I will be returning to these issues again in the following sections. Those pigs which were engineered with a human growth hormone grew slightly faster than normal, but they were arthritic, had ulcers and were partially blind. This was caused by excess production of growth hormone. In other words, although the gene could be introduced into the pig genome, there was no regulation of the expression of the gene. This example illustrates one of the risks of transgenic manipulation, that there is no way of knowing in advance how far it will be possible to regulate such genes. Furthermore, even if the regulation is achieved in short-term 'laboratory' experiments, there is no guarantee that this will apply in the field. If we apply this concept to plants containing genes from a genetically engineered bacteria, the potential threats are even more worrying. What could safely control bacteria released accidentally into the environment?

There are many other examples which I could use to show how biotechnology can become a threat to the environment, animal life, human health and genuine human existence. The close relationship between farmers and their land seems to be replaced by the transference of power to seed companies. I will describe one more example to illustrate this point. In 1992 a biotechnology subsidiary of W. R. Grace (USA), called Agracetus, was given the patent on all genetically engineered cotton plants.[21] The patent covers any genetically engineered cotton species, which gives the company a monopoly on all newly developed strains. Agracetus can thus charge royalties to any company or scientist intending to genetically engineer this crop. In effect this company has absolute power on cotton growing. Even the biotechnology industry regarded this as an unfortunate anomaly. Yet the European patent office has repeated this same mistake in giving Agracetus a patent on all genetically engineered soya plants. Those most likely to suffer in this case are the Third World farmers who cannot afford to pay the high licence fees. The Rural Advancement Fund (International) helps countries in the Third World to monitor changes in biotechnology. It has declared its intention to challenge the soya patent through legal means.

The potential threat of genetic engineering seems to have been anticipated by Pope John Paul II, who declared in his address to UNESCO:

> The future of mankind is threatened, radically threatened, in spite of very noble intentions, by men of science . . . their discoveries have been and continue to be exploited – to the prejudice of ethical imperatives – to ends of destruction and death to a degree never before attained, causing unimaginable ravages. This can be verified as well in the realm of genetic manipulations and biological experiments as well as those of chemical, bacteriological or nuclear armaments.[22]

While his direct accusation of scientists alone seems to my mind to be misplaced, it is true to say that the potential abuse of the power of genetic engineering represents an enormous threat to the survival and quality of life of humans, animals and plants alike. Adequate restraints and controls cannot, therefore, be left just to government commissions or companies exercising their own internal monitoring.

A philosophical critique

The French philosopher and social critic Jacques Ellul describes technology as an autonomous and uncontrollable force which pervades social, economic and political life.[23] This leads to an enslavement to all that the technology demands. If we extend this idea to genetic engineering, then the very fabric of life becomes subject to a form of determinism. At the opposite extreme, it is possible to portray technology as a liberator, a product of human choices. Samuel Florman is an engineer who argues that the life of earlier centuries has been overromanticized.[24] He believes that the undesirable effects of technology can be overcome by more technology. One example of this would be the use of genetic engineering to overcome the problem of crop sensitivity to herbicides.

The extent to which we perceive genetic engineering as a threat or a promise reflects a wider human dualism in our own perceptions. On the one hand we are actively involved in our individual schemes and projects, while on the other hand we stand apart from these and adopt a more holistic perspective.[25] The beneficial effects of genetic engineering in the development of new medicines and the protection of animals and plants, and in some cases humans, against disease is often cited by genetic engineers to justify their work and achievements. However, the root cause of the disease may be overlooked. For example, many of the diseases in animal husbandry are fostered by overcrowding and other unhealthy conditions. To engineer the animals against disease does not get at the root of the problem. The unashamedly anthropocentric philosophy which is behind these developments is anthropocentrism at its worst: namely use of a particular technique purely for individual commercial benefit. On the other hand, the wider impact of these developments and the very idea of interfering with genetic material in an irreversible way gives rise to genuine concern that genetic engineering is to the detriment of life and the planet as a whole.[26] A lack of holistic thinking is the product of traditional scientific method. A reductionist methodology in science analyses the separate components as a way of understanding life. Genetic engineering is inevitably the fruit of this reductionist approach. A more holistic approach within biology is taken by ecologists, so that a radically different method leads to conflict even amongst scientists as to the goals of their research. It seems to me that this ambiguity mirrors that felt by the wider public in relation to the value or otherwise of genetic engineering.

It might be possible to argue against any genetic engineering on the basis that it is somehow unnatural. However, it is important here to distinguish between the use of genetic engineering to speed up what would be possible in normal breeding methods and its use in transgenic experiments. There are those who object to genetic engineering on the basis that it is an invasion of biological integrity. However, according to biologists' understanding of evolution, biological integrity as such does not exist. As a consequence, we share many of the biological and physiological processes as other life forms. Furthermore, the ability of breeders and farmers to bring about change over a relatively short time span became part of the evidence which has been used to support Charles Darwin's theory of evolution.[27] It seems questionable whether natural selection is any more altruistic for the species compared with artificial breeding or genetic engineering. It seems to me that the philosophical basis for regulation and constraints in genetic engineering needs to be sought in avenues other than a vague notion of biological integrity. I will return to this point again below.

In all breeding methods there is a tendency to treat animals and plants as commodities. Genetic engineering allows, then, an even greater detachment from the animal or plant in such a way that they can become highly vulnerable to exploitation. Heidegger rejects the idea that technology is neutral and is simply a means to an end.[28] He suggests that modern technology has failed to bring forth what is the original intention of the natural environment. Instead technology is confrontational and challenging. As such, this is an unreasonable demand placed on the natural world. Heidegger argues that this attitude in technology was prior to the modern physics of the seventeenth century – and I could add prior to the modern biology of the twentieth and twenty-first centuries. He believes that the greatest illusion for human beings is to see everything as their own construction, since it drives out other forms of revealing from within the natural world. Heidegger did not exactly reject technology, but was acutely aware of its ambiguities for a genuinely human existence. The same could be said for the biotechnology revolution of contemporary Western culture.

As I discussed in the last chapter, Paul Taylor has suggested that respect for the natural world is a key paradigm in the development of a theory of environmental ethics. We are all part of a single biotic community, a point taken up and developed by philosophers such as Baird Callicott. Whereas Taylor argues for the individual inherent worth of all living creatures, Callicott argues for the inherent value placed on the whole by the human subjects.[29] The key question is whether a recognition of mutual dependence constitutes a moral relationship as well. Robin Attfield argues against the idea that interdependence strengthens moral relationships, preferring the idea that all species who have interests have moral standing.[30] He believes that 'rights' are not the only basis of moral concern; something can lack 'rights' but still have moral standing. I could ask, what does it mean for a species to have interests? This seems to be related to the idea of what constitutes respect. Kant believed that if we treat people as a means to our own ends and do not recognize their ends, we are failing to show respect.[31] Genetic engineering has to treat living things in a mechanistic way in order to achieve its goals. However, there is a distinction between treating a living thing purely as a means for our own ends, disregarding the creature's ends, and bringing our interests into line with that of the creature. This echoes the idea of Heidegger that we need to become sensitive to the 'revealing' within the natural world. However, it still requires human judgement and a form of empathy to decide exactly what the interests of the creature mean in practice.

There seems to be no need to reject all genetic engineering in principle, as long as we take into account the interests of the creatures concerned. In the case studies cited above, the deliberate manipulation of animal hereditary material as a means of rendering the animal more passive in crowded conditions is just one example among many of an unacceptable violation of that creature's interests. It is also clear that the extent of genetic manipulation that is acceptable at the level of the creature's interest must be qualified by reference to the evolutionary complexity of an organism, as suggested by van de Veer's two factor egalitarianism. However, I am concerned that we do not give blanket approval to all transgenic manipulations of plants and other so-called 'lower' organisms, such as viruses and bacteria. Where the manipulation is driven by the need for commercial gain of a limited number of human individuals, it is surely against the principle of respect for all life.

It seems to me that while the idea of respect for life and the interests of creatures can take us some way towards working out priorities in genetic engineering, if we just focus on the immediate interests of the creature, we can all too easily lose sight of the wider social and environmental consequences. I would also add that we need to take into account the long-term interests of the environment as a whole, as well as the interests of the whole human community. For example, if I take the example of herbicide resistance introduced by genetic engineering, there seems to be little evidence that this causes immediate harm to the species involved. If anything, the crop benefits seemingly from the change as now it is resistant to herbicides. Nonetheless, as I showed above, the potential effects on the ecosystem and the farming community could be catastrophic. In this case and in other examples I have cited, the interest in profit seems to be higher than respect for the wider community.

As I discussed in Chapter 3, Holmes Rolston III has argued for the idea of systemic value as a way of taking into account the worth of the whole ecosystem(s). This could possibly be a useful concept as applied to the example I have mentioned above. However, the idea needs some qualification, as it can lead to an overromanticized view of the biological integrity of the system, which is itself a highly debatable topic amongst ecologists. Ecosystems emerge in a more random way than is implied by some 'deep green' philosophers. Having said this, there is no guarantee that the new ecosystem that would develop after human interference would be either desirable or controllable. I will return to this point again below. It is ironical, perhaps, that genetic engineering, which seeks to assert human power over the natural environment, can lead to situations which could, potentially, become uncontrollable. The Utopian dream of a custom-made world is supposed to lead to a fully controlled environment for human habitation. Charges of sentimentality abound, both against genetic engineers who have held such a vision, and in return, against the animal liberationists. As I mentioned in Chapter 3, animal rights activists, such as Tom Regan, press for the abolition of all commercial agriculture, which in effect amounts to a virtual moratorium on genetic engineering of animals.

The philosopher A. A. Brennan has argued that one of the main problems in making decisions about the environment is our lack of honesty: 'It is neither unfair nor unkind to governments, public agencies and corporations to observe that we are a long way from full honesty in our debates and deliberations on the environment.'[32] He insists that this dishonesty is encouraged by the public acceptance of a shallow analysis of science and mythic portrayals of our situation, both of which encourage 'self-deception' and 'incontinence'. The first myth he highlights is that of restoring nature, after human interventions such as mining, industrialization, etc. There is a strong belief that, given the right technology, we could restore nature to the original condition. I could add here the myth of improving nature as applied to genetic engineering. A good example would be the attempt by scientists, so far unsuccessful, to transfer nitrogen-fixing genes from legumes, such as clover and peas, to cereal plants, such as wheat. Such transgenic experiments promise to improve nature by giving wheat plants the potential to fix gaseous nitrogen so that they would become less dependent on artificial fertilizers.[33]

In the 'natural' state legumes fix nitrogen by relying on a bacteria which occupies its host plant in special swollen parts of the roots called nodules. Although the project sounds altruistic and environmentally friendly, the dishonesty lies in a failure to point out physiological and ecological features. Firstly, even legumes rely on artificial fertilizers to increase yields, so that it is by no means certain that such manipulation would reduce fertilizer use sufficiently to prevent environmental damage. Secondly, and perhaps more importantly, the technique involves modifying the bacteria–host relationship in such a way that the bacteria is no longer host specific. There seems to be no guarantee that such a relationship would be stable in field conditions, raising the spectre of the release of modified bacteria into the environment. The dishonesty lies in a failure to recognize that the project is not as environmentally friendly or sustainable as it appears. More often than not such projects are given an environmental gloss as a way of appropriating funds for the patent of modified seeds or, more simply, as a way of enhancing the publication record of an institution, thus giving access to yet greater power and influence in a particular area of research.

Brennan also asks the question as to whether the restoration of the natural environment after industrialization would give the northern continents more authority in demanding that southern countries conserve their rain forests. However, it is biologically naive to assume, even in the forests of the north, that any real restoration to the original diversity is possible. In the tropics the rate of elimination of species is so high that all talk of future restoration is wishful thinking. Yet genetic engineers, predominantly in richer northern nations, are relying on the Third World as a reserve for potential genetic variability.

Another common myth is that of 'wild' nature. Holmes Rolston III has used this idea as a paradigm for his philosophy of environmental ethics.[34] Attached to this myth is the concept that all 'wild' ecosystems are both stable and diverse. While the characteristic of biodiversity is true for the tropical rainforest, it is not true for all other ecosystems. For example, Horn has shown that the New England woodland has a sequence of regrowth, known as succession, towards the reduction of biodiversity.[35] I am not saying that the preservation of biodiversity is mistaken, rather it cannot be supported by reference to 'wild' nature. I would agree with Brennan that: 'It is striking and unfortunate that many conservationists still operate with ideas of balance and diversity in nature that were more prevalent in the nineteenth century than among contemporary ecologists.'[36]

The myth of balance and diversity is important for 'deep green' philosophers as it seems to provide a biological basis for non-interference. However, absolute non-interference is not really an option for humans, any more than any other species. It is the form of meddling that raises moral, aesthetic and policy issues. The biodiversity of species in this context needs to be carefully distinguished from the variability in a given species that I mentioned above in relation to genetic engineering. This natural variability in one species is an in-built mechanism for protection against disease. The loss of species themselves through loss of highly diverse ecosystems, as in the tropical rainforests, cannot be desirable. It seems to me that Brennan is correct in his belief that the basis for the maintenance of diversity cannot be sought through a form of 'naturalism' which exalts a selective and distorted view of ecology. In other words, Rolston's thesis seems to imply that it is only 'wild' nature that has value. I think that Attfield is correct to point out that there is a case for the careful cultivation of national parks and other 'artificial' natural systems as places with inherent value, in addition to a simple preservation of 'wilderness', which in Europe at least is virtually non-existent. While I argue the case against some of the extremes of genetic engineering practice as applied to agriculture, I believe that humanity still has a responsibility to work with the natural world in creative cultivation and sustainable development. I will return to the issue of sustainability in the next chapter. As Attfield points out:

> Besides the role of rehabilitating nature, people have the role of making nature habitable, and sustainably so – habitable for human beings of the present and the future and also for those non-human creatures which they rear or cultivate as part of their civilization or culture – without at the same time undermining those whose wildness and otherness they need or whose flourishing is of value in itself.[37]

The protection of the interests of the species seems to be a more fruitful approach as long as it is set in the context of the interests of global ecology and issues of justice related to the human community. There are other arguments which I could use which overlap with theological perspectives, outlined below.

To conclude this section: there seems to be no real philosophical basis for complete abstinence towards genetic engineering as applied to agriculture. Rather, following Heidegger, it seems to me that we need to work towards the transformation of genetic engineering so that it comes to represent a more fully humane enterprise, in touch with the immediate and long-term effects.

A theological critique

William Frankena, in his analysis of the potential of theology for ethics, argues that while theology does have an 'ethic', it cannot answer all questions by itself.[38] The problems of interpretation and application still leave difficult decisions which cannot be answered by theology alone. The normative elements in biblical theology, such as the Ten Commandments, may be a guide for living, but this is different from morality as such. For example, the sabbath day commandment is less a moral question than one of lifestyle. A more relevant contribution of theology, perhaps, in the present context, is the way theological perspectives influence ethics. Frankena argues that logically ethics need not rest on theological presuppositions. However, the latter do show us what areas are particularly important to consider in ethics and provide both a rationality and motivation for ethical inquiry. Frankena's position is intermediate between those who argue that theology is impotent for ethics, in other words that ethics is autonomous, and those who use theology as a basis and foundation for ethics.

The concept of autonomy in ethics seems to foster a broadly Kantian version of the division of labour for ethical decision making, where ethics becomes a disembodied autonomous subject legislating for itself on the basis of disengaged reason alone.[39] There is a tendency for this model to collapse into the idea that science alone can solve all environmental problems, since ethical value becomes objectified and unambiguous. An example, in practice, of the way science has attempted to introduce a notion of value as part of its directives is the practice of cost-benefit and risk-benefit analysis. In this case the sum preferences of all individuals affected by a decision are taken into account in arriving at 'ethical' judgements.[40] The question still remains: how ethical are such practices? Can we really put a price on the value of a natural habitat? Who are the beneficiaries really? What seems to happen more often is risk-benefit is calculated according to what is legal, rather than what is valuable.

It seems to me that any claim of autonomy on the part of ethics is naive in its failure to recognize the complex cultural context, which includes a religious perspective. Furthermore, such an approach assumes, incorrectly, that all ethics would necessarily come to a univocal position in a way which ignores the concrete realities of ethical dilemmas. While environmental pragmatism, which I discussed in the last chapter, attempts to come to terms with diversity, it seems to assume that the context itself will set the parameters for conflict resolution. For McCormick the biblical story of faith becomes the 'overarching foundation and criterion for morality'.[41] A Christian speaks out of the experience of this story, so that reasoning in ethics is informed by faith, and theology 'yields a value judgement and a general framework or attitude. It provides a context for subsequent moral reasoning'.[42] McCormick insists that we truncate the task of theology if we see it as an action guide, rather than looking at wider issues such as the quality (good or bad) of the agent. This does not mean that all morally relevant insights are specifically Christian; rather, a Christian perspective confirms and critiques ethical practice.

For theology to be true to its task it must include a reference to an ultimate power, a theocentric ethic which I outlined in the last chapter.[43]

Just as it is inappropriate to use religion to sustain moral causes and purposes, which renders the deity merely incidental to human ambitions, so it is unjustified to incorporate values into policy making as a means of influencing particular social action. This could apply in the example of genetic engineering of nitrogen fixation that I mentioned above. Purported environmental 'values' were incorporated into the project as a means of gaining access to power, rather like the biblical account of the wolf dressed up in sheep's clothing. In other cases, so-called values can be used just as a means of influencing public opinion. The rationale for this move is that bare facts alone have failed to have any real influence, so values, including religious ones, are introduced as a way of encouraging particular forms of behaviour.[44] The goals, then, are set prior to any consideration of theology or ethics and values are introduced so as to act as a 'social lubricant'. This form of social engineering represents an unacceptable use of theological ethics.

If we start with the social issues of our time there is a danger that religion will somehow become fitted in, put to the use of the immediate needs of groups and individuals. In this sense God becomes an instrument who endorses human action. This is not my intention. Rather, by beginning with some of the issues associated with genetic engineering in agriculture, the theological and ethical discussion becomes more rooted in the real problems we are facing. The opposite danger, namely a theological discussion which never moves further than rarefied concepts of God, seems to me to be more prevalent amongst contemporary theologies of creation, which I outlined briefly in Chapter 2. The failure of science alone to come up with any effective answers to complex environmental issues is occasion enough to widen the debate to include other disciplines. This is especially true as applied to biotechnology where it is still possible for theologians, philosophers and ethicists to contribute to the shape of this rapidly expanding field.

The question now is: how can the language about God help to shape the ethical directives in genetic engineering? For some religious believers the very idea of genetic engineering sounds like blasphemy. What right have humans to interfere with the natural world when it has been declared by God as good? In answer to this question I refer to the above discussion about the interference of humans in 'wild' nature through the cultivation of land since the dawn of human existence. The questions are: what kind of interference is justified, given a belief in the goodness of creation?; what are the social and environmental consequences? The current discussion about the morality of genetic engineering has focused almost exclusively on medical ethics and the human genome. Lehmann, for example, argues that we should suspend genetic engineering because it goes against the principle that 'knowledge of life processes must be used to reinforce what is human in us'.[45] In other words, genetic engineering is somehow dehumanizing. According to biblical anthropology human beings are made in the 'image of God', so that to interfere with human genetic material might seem to go against the special position of humans as made in the divine image.

How far can we extend this idea of the divine image in order to apply a moratorium on the genetic engineering of animals and plants? The theological paradigm that seems relevant here is the experience in the Judaeo-Christian community of a God of love. The command of God in Genesis for humans to exercise dominion over creation is qualified by the essence of the relationship between Creator and creation as one of loving involvement. Creation becomes an expression of the love and glory of God. As Moltmann points out:

> The world as a free creation cannot be a necessary unfolding of God, nor an emanation of his being from divine fullness. When he creates something that is not God but also not nothing, this must have its ground not in itself, but in God's will or pleasure. It is the realm in which God displays his glory.[46]

The new creation will be one where God will be 'all in all'. The world becomes transfigured by the presence of God through the participation of creation in God's infinite creativity.[47] This echoes the Eastern Orthodox concept of participation creation in the energia of God.[48] Karl Rahner, similarly, insists that it is God's intention to give creation a supernatural end, which has an effect on the essence of being itself.[49] For Rahner, the natural knowledge of God as perceived in creation is not sharply distinguished from the revelation of God. However, the Christological dimension in Rahner's thought qualifies the themes of the future glory of creation. Moltmann, similarly, insists that the cross of Christ reminds us that the future of creation is not Utopia on earth.

Given this theological perspective which stresses the love of God and the future glory of creation, how does this effect environmental ethics? First the cross of Christ reminds us both of the reality of the suffering of creation and the very real temptation of humans to sin in identifying their human enterprise with absolute value. Genetic engineering can never achieve Utopia on earth, especially when we are blind to its use as an instrument in suffering. Moreover, God's love for all creation demands a respect for the interests of all creatures, whether they are produced by genetic engineering or traditional breeding methods. This applies, as well, in making policy decisions as to the direction of research. Human beings, as made in the image of God, share in the creativity of the Creator. The Creator's intention is towards future glorification. As I argued in the last chapter, we can hope for the final glorification of all creatures, not just humanity. Hence the human motivation to develop new varieties and transgenic species needs to be carefully scrutinized. The possibility of dishonesty in the use of environmental language to cover up materialist or power-craving instincts has to be exposed. In this scenario environmental ethics overlap with business ethics and social conduct.

The latter raises questions about human justice. For some years questions to do with development were considered in a way that was detached from environmental issues. Development workers tended to despise environmentalists as those who seemed to pay more attention to the survival of animals, rather than people. More recently, there has been a greater appreciation of the interrelationship between environmental problems and development issues. It seems to me that this linkage is of special relevance to the particular questions surrounding genetic engineering, as highlighted in the case studies above. A Christian theological perspective would insist on examining the long-term consequences to poorer nations and communities.[50] It is this broader view which is essential to keep in mind when dealing with decisions about the validity of particular genetic engineering projects. The wider global environmental impact also has to become part of the consideration by those given the power to implement policies. Hence it is a question both of what is good for humanity as well as what is good for the whole natural environment. In this way the wider human and cosmic contexts act as twin points of reference. A theocentric perspective avoids the difficulty of somehow combining both anthropocentric and holistic perspectives, which seems to be advocated by philosophers such as Lawrence Johnson and environmental pragmatists discussed in the last chapter. It seems to me that such a median view still leaves the question of the conflict between those who advocate intrinsic value of the whole and those who argue for the priority of human interests unanswered. A theocentric perspective, by keeping the dignity of humanity but relativizing all human achievements from the perspective of a God who loves all creation, can give us a guide through some of the dilemmas.

A theological critique requires a radical change of attitude in formulating the goals of genetic engineering from one based on consumerism and the individual pursuit of happiness, to a more community based view which includes respect for the whole environment. This respect is understood in terms of knowledge of the Creator who created the world out of nothing. Some theologians have attached the idea of respect for the natural environment to the concept of Gaia. As I argued in the last chapter, basing an environmental philosophy on Lovelock's Gaia hypothesis has its own pitfalls scientifically, theologically and ethically.

For some theologians, even the idea of genetic engineering represents an unacceptable expression of human power and domination over creation. Linzey, for example, believes that: 'Genetic engineering represents the concretization of the absolute claim that animals belong to and exist for us . . . *they become totally and completely human property*.'[51] He describes genetic engineering as a form of animal slavery. The idea that animals and plants are made simply for human use has its roots in the teaching of Aristotle.[52] A similar view is echoed by Thomas Aquinas, who insisted that non-human creation was for the benefit of humans.[53] He distinguishes between the general care for animals and the love and fellowship among the human community.[54] The idea of cosmic fellowship and friendship with the whole creation, to which I alluded in the last chapter, has become a popular motif amongst contemporary theologians concerned with the environment.[55] While I would argue against Aquinas's notion that all creation exists purely for human benefit, there seems to be some validity in distinguishing the love among humans from the love of humans for all of creation. As I tried to show earlier, if we deny that there are any distinctions we end up with a total moratorium on all genetic engineering and probably many forms of traditional agriculture. This seems to be the logic of extending rights to animals and leads Linzey to declare genetic engineering in these terms, that 'Nothing less than the dismantling of this science as an institution can satisfy those who advocate moral justice for animals.'[56] Yet even in the human community some genetic engineering is deemed acceptable, especially in the somatic line rather than the germ line. Somatic refers to the cells of an individual body, while the germ line is the hereditary material passed on to the next generation. Now if we apply this principle to plants, which are totipotent, that is any cell can be used, in theory, to generate another plant, the strictest rules governing genetic engineering are a nonsense. It seems to me that there need to be much tighter controls on transgenic experiments, which are not possible by conventional breeding, but to ban all genetic engineering is unrealistic and most likely to be dismissed by those who are already in power.

It would be inappropriate to lay the blame for the abuse of genetic engineering on the scientists alone. We are all implicated in the social web of which scientists are a part. In seeking for a change in attitude amongst those more directly involved, a wider *metanoia* is needed which incorporates a sensitivity to creation in every aspect of our lives. This metanoia includes an attitude of humility and respect for all members of the human and non-human community. It seems to me that we cannot avoid sacrificial effort on our part. In the words of the Ecumenical Patriarchate:

> This is a new situation, a new challenge. It calls for humanity to bear some of the pain of creation as well as to enjoy and celebrate it. It calls first and foremost for repentance, but of an order not previously understood by many.[57]

If we love creation we begin, then, to see the 'divine mystery in things'. This awareness of the sacred in creation can become the lens through which we seek out our responsibilities in caring for the earth. The most common mind-set amongst genetic engineers is to fix on a particular problem or goal and then to find ways of achieving this goal. A theological approach encourages those who are involved to see the wider social and religious consequences of these decisions. It does not necessarily ban all genetic engineering, but seeks to transform it so that it more fully represents a humane enterprise.

Philip Sharrard's book *Human Image: World Image* is an important restatement of creation as sacred.[58] However, he is far too harsh in his treatment of modern science and scientists. He believes that any attempt to interfere or remodel creation amounts to 'sheer folly', treating scientists as scapegoats, especially the mathematicians. Such a move is not particularly helpful. It seems to me that understanding the world as in some sense sacred is compatible with science as long as it seeks ethical ways forward. A complete retreat from science is a refusal to face the realities of this world. Sherrard's rejection of all dualistic thinking, which he describes as a 'form of depravity' seems to extend to his understanding of the relationship between God and the world. As we might expect, he rejects the concept of creation of the world out of nothing, and in doing so blurs the distinction between God and creation.

The ethical consequences of a fully fledged antidualism are somewhat disturbing. Szerszynski has summed up the arguments against antidualism.[59] One of his most important insights is that a refusal to distinguish between ourselves and creation leads to a narcissistic identification with the earth. For Matthew Fox and others, sin becomes treating others as separate from ourselves.[60] Instead of a genuinely relational ethic we arrive at an incorporation of the other into ourselves. Szerszynski points to the increasing tendency to turn to the non-European traditions for a saving role in ecology 'as if they were the repressed contents of the European unconscious'. However, there seems to be no link between cosmologies and potential damage to the natural environment.[61] As I mentioned in the last chapter, environmental ethics necessarily derives from the modern understanding of the natural environment as other. If we pretend that this is not the case, we end up in self-delusion.[62] While I think that Milbank makes a good point, I do not agree that it follows from this that an understanding of the world as sacred is always misplaced. A true understanding of the sacred respects the difference in the other. Moreover, the sacred is not the same as the divine. That which is sacred is not an object for our worship, but a living creature with which we enter into a respectful relationship. Like the Celtic saints, a keen awareness of the transcendence of God is matched by an appreciation of God's immanence in creation.[63]

Hans Küng's book *Global Responsibility* is a brave attempt to locate world religions in the field of ethics.[64] However, his suggestion that a single world ethic can emerge from many different religious perspectives seems to me to be misplaced. It would tend to deny the distinctive contribution of each culture. There is also disappointingly little reference in his work to the global problems of ecology. While the dialogue between different faiths is important in contributing to the practical task of policy making in genetic engineering, each faith needs to become aware of its own distinctive contribution. A lowest common denominator approach is unlikely to yield the promised fruit and more likely to raise false hopes.

Conclusions

The revolutionary changes taking place within biological science, in particular genetic engineering, have encouraged a mixed sense of awe and threat. Lewontin has suggested that the power of DNA technology is such so as to give biology the status of 'ideology'.[65] I have argued in this chapter that it is application of genetic engineering in biotechnology that raises significant philosophical, ethical and theological questions. The awe at the dawn of the new technology is tempered by the potential of this technology to be abused as a means of power and domination. While on the one hand the technology promises solutions to immediate technological and scientific problems, on the other hand the indirect effects of such technology can easily be overlooked by policy makers interested only in immediate, short-term damage limitation. A philosophical perspective can offer a more holistic approach. However, the language of philosophy can be out of touch with modern developments in biology. Various myths abound, such as the goodness of 'wild nature' or its integrity. A real transformation of genetic engineering which would allow it to become a more fully humane enterprise is rarely discussed.

A theological approach offers a spiritual dimension to the discussion and asks, in particular, how these changes impinge on those who are part of the Christian story. It challenges the injustices raised in the human community, as well as the more indirect damage to the environment. It seeks for caution in any genetic engineering which involves transgenic manipulations. It also refuses to concentrate on the sacred in nature without considering the difference between God and creation, humanity and other life forms. If we refuse to acknowledge these differences, or if we return to a pantheistic vision of creation, science and genetic engineering *per se* become impossible. Such an approach seems to be out of touch with modern biology, a romanticism which is not helpful in post-Enlightenment culture. It is, furthermore, the very opposite of what it means to be holistic. It seems to me that an understanding of God as transcendent, yet immanent in creation, provides a way through the current vogue of 'naturalism', into a more fruitful and realistic dialogue with science.

NOTES

1. This chapter is a revised, expanded and updated version of an article published in *The Hethrop Journal*, vol. 36, no. 3 (July 1995), pp. 307–27.
2. See, for example, M. Midgley, *The Ethical Primate: Humans, Freedom and Morality* (Routledge, London, 1994); World Council of Churches, *Manipulating Life: Ethical Issues in Genetic Engineering* (WCC, Geneva, 1982); J. Mahoney, *Bioethics and Belief* (Sheed and Ward, London, 1984); A. C. Varga, *The Main Issues in Bioethics*, rev. edn (Paulist Press, Dublin, 1984); J. Harris, *Wonderwoman and Superman: The Ethics of Biotechnology* (Oxford University Press, Oxford, 1994); D. Suzuki and P. Knudtson, *Genethics: The Ethics of Engineering Life* (Unwin Hyman, London, 1990).
3. W. Temple, *Christianity and Social Order* (Penguin, London, 1942).
4. S. Clark, *How to Think About the Earth* (Mowbray, London, 1993), p. vii.
5. I. Barbour, *Ethics in an Age of Technology* (SCM Press, London, 1992).
6. For a more general discussion of the relationship between theology and biology, see C. Deane-Drummond, 'Biology and Theology in Conversation: Reflections on Ecological Theology', *New Blackfriars*, vol. 74 (October 1993), pp. 465–73.
7. S. Wortmann and R. Cummings, *To Feed the World* (John Hopkins University Press, Baltimore, 1978).
8. M. Lipton and R. Longhurst, *New Seeds and Poor People* (John Hopkins University Press, Baltimore, 1989).
9. E. C. Wolf, *Reversing Africa's Decline* (World Watch Institute, Washington, 1985); E. C. Wolf, *Beyond the Green Revolution: New Approaches to Third World Agriculture* (World Watch Institute, Washington, 1986).
10. I. Barbour, *op. cit.*, pp. 191–2. See also P. Wheale and R. McNally, *Genetic Engineering: Catastrophe or Utopia?* (St Martin's Press, New York, 1988); P. Wheale and R. McNally (eds), *The Bio-Revolution* (Pluto Press, London, 1990).
11. D. King, 'The FLAV SAVR Tomato – Hard Tomatoes, Hard Times', *GenEthics News*, issue 3 (September/October 1994), p. 8; D. King, 'Supermarkets to Label Engineered Tomato Paste', *GenEthics News*, issue 8 (September/October 1995), p. 4. Note that the journal *GenEthics News*, launched in 1994, gives up-to-date information on all the latest releases of genetically engineered organisms.
12. Frost resistance in strawberries is achieved by genetic engineering of a bacterium that normally lives on the strawberries' surface and acts as a nucleation site for ice crystals. The modified bacterium loses this capacity. See K. Dahlberg, *New Directives for Agriculture and Agricultural Research* (Rowman and Allanheld, Lanham, 1986).

13. For a series of articles on bovine growth hormone (BGH), otherwise known as bovine somatotropin (BST), see P. Wheale and R. McNally (eds), *The Bio-Revolution*, *op. cit.*, pp. 57–100; I. Barbour, *op. cit.*, pp. 192–4.
14. See especially A. Holland, 'The Biotic Community: A Philosophical Critique of Genetic Engineering' in P. Wheale and R. McNally (eds), *ibid.*, pp. 171–2.
15. D. King, 'Government Allows Unlimited Release of Genetically Engineered Plant', *GenEthics News*, issue 1 (1994), p. 3.
16. *Ibid.*, p. 2.
17. D. King, 'Should We Tolerate Herbicide Tolerant Plants?', *GenEthics News*, issue 5 (1995), p. 7.
18. See C. Juma, *The Gene Hunters: Biotechnology and the Scramble for Seeds* (Zed Books, London, 1989).
19. D. King, 'Genetically Engineered Food: How Safe is it?', *GenEthics News*, issue 9 (November/December 1995), p. 8.
20. D. King, 'Animals, Genes and Ethics', *GenEthics News*, issue 2 (July/August 1994), p. 8; D. King, 'And How the Cow Jumped Over the Moon', *GenEthics News*, issue 3 (September/October 1994), pp. 6–7.
21. D. King, 'Patent Offices Hand Cotton and Soya Monopolies to W. R. Grace', *GenEthics News*, issue 1 (1994), p. 3.
22. Cited in G. J. V. Nossal and R. L. Coppel, *Re-Shaping Life: Key Issues in Genetic Engineering*, 2nd edn (Melbourne University Press, Melbourne, 1989), pp. 132–3. The authors believe that statements such as these just frighten people and foster disillusionment with science. They retort that it is 'unfair to blame science and technology for the ills in the human condition that are as old as mankind', p. 134.
23. J. Ellul, *The Technological Society*, trans. J. Wilkinson (Knopf, New York, 1964); J. Ellul, *The Technological System*, trans. J. Neugroschel (Continuum, New York, 1980); J. Ellul, *The Technological Bluff*, trans. G. Bromiley (Eerdmans, London, 1990).
24. S. Florman, *Blaming Technology: The Irrational Search for Scapegoats* (St Martin's Press, New York, 1981).
25. See A. Holland, *op. cit.*, pp. 166–74.
26. *Ibid.*, pp. 166–74. See also chapter 12, 'Maize: In Praise of Genetic Diversity', in D. Suzuki and P. Knudtson, *Genethics: The Ethics of Engineering Life*, *op. cit.*, pp. 276–303.
27. P. Wheale and R. McNally, *Genetic Engineering*, p. 19.
28. M. Heidegger, *The Question Concerning Technology and Other Essays*, trans. W. Lovitt (Harper Torchbooks, London, 1969), pp. 3–35.
29. See C. Palmer, 'A Biographical Essay on Environmental Ethics', *Studies in Christian Ethics*, vol. 7, issue 1 (1994), pp. 68–97.
30. For Attfield moral standing is given to a living thing that has the ability to flourish and develop. This ability to flourish gives the organism interests. Attfield adopts a broadly consequentialist view, ascribing value to a state of being of a living organism, rather than the organism itself. This contrasts with Paul Taylor's deontological approach, which seems to affirm biocentric equality of value of all organisms in and of themselves. R. Attfield, *The Ethics of Environmental Concern* (Blackwell, Oxford, 1983), pp. 158 ff.; P. Taylor, *Respect for Nature: A Theory of Environmental Ethics* (Princetown University Press, Princeton, New Jersey, 1986). For a comparison of different perspectives, see C. Palmer, *op. cit.*, pp. 68–97.
31. I. Kant, *Critique of Practical Reason* (London, 1909), p. 47.
32. A. A. Brennan, 'Environmental Decision Making', in R. J. Berry (ed.), *Environmental Dilemmas: Ethics and Decisions* (Chapman and Hall, London, 1992), p. 18.
33. For an excellent introductory account of the biology of nitrogen fixation, see J. I. Sprent, *The Biology of Nitrogen Fixing Organisms* (McGraw Hill, New York, 1979). For a more detailed monograph which includes the genetics of nitrogen fixation, see J. Postgate, *The Fundamentals of Nitrogen Fixation* (Cambridge University Press, Cambridge, 1982).
34. H. Rolston III, *Philosophy Gone Wild: Environmental Ethics* (Prometheus Books, Buffalo, 1989).
35. H. S. Horn, 'Markovian Properties of Forest Succession', in *Ecology and the Evolution of Communities*, M. L. Cody and J. M. Diamond (eds) (Belknap Press, Harvard University, Cambridge, 1975), pp. 196–211. While many 'patterns' have been found in communities of species, there is no evidence for fixed relationships which would suggest an 'organismic' model for community ecology. Most ecologists accept the idea of 'limited membership', that

is, there are restrictions to the range of species found in any given area depending on the physical conditions, dispersal of seeds and interactions between species. See J. Roughgarden and J. Diamond, 'Overview: The Role of Species Interactions in Community Ecology', in *Community Ecology*, J. Diamond and T. J. Case (eds) (Harper and Row, San Francisco, 1986).

36. A. Brennan, *op. cit.*, pp. 16–17.
37. R. Attfield, 'Rehabilitating Nature and Making Nature Habitable', in *Philosophy and the Natural Environment*, R. Attfield and A. Belsey (eds) (Cambridge University Press, Cambridge, 1994), p. 55.
38. W. K. Frankena, 'The Potential of Theology for Ethics', in *Theology and Bioethics*, E. E. Shelp (ed.) (D. Reidel Publishing, Dordrecht, 1985), pp. 49–64.
39. R. Grove-White and B. Szerszynski, 'Getting Behind Environmental Ethics', *Environmental Values*, vol. 1 (1992), pp. 285–96.
40. See, for example, D. Pearce, A. Markandya and E. B. Barbier, *Blueprint for a Green Economy* (Earthscan, London, 1989), pp. 56–7.
41. R. A. McCormick, 'Theology and Ethics: Christian Foundations', in E. E. Shelp (ed.), *op. cit.*, p. 96.
42. *Ibid.*, p. 98.
43. See also J. M. Gustafson, *The Contribution of Theology to Medical Ethics* (Marquette University Press, Milwaukee, 1975); J. M. Gustafson, *Theology and Ethics* (Blackwell, Oxford, 1981).
44. R. Grove-White and B. Szerszynski, *op. cit.*, pp. 289–90.
45. P. Lehmann, 'Responsibility for Life: Bioethics in Theological Perspective' in E. E. Shelp (ed.), *op. cit.*, p. 294. Most scholars recognize that there is a difference ethically between genetic modification of the cells of the body, known as the somatic line, and the hereditary material, known as the germ line. Modification of the latter is less acceptable compared with the former. For a discussion see J. Harris, *op. cit.*, pp. 162 ff.; D. Suzuki and P. Knudtson, *op. cit.*, pp. 163–191.
46. J. Moltmann, *Theology and Joy*, trans. R. Ulrich (SCM Press, London, 1973), p. 41.
47. See J. Moltmann, *Religion, Revolution and the Future*, trans. M. D. Meeks (Charles Scribner's Sons, New York, 1969), p. 34. This theme recurs in his subsequent works. For a discussion, see C. Deane-Drummond, *Towards a Green Theology Through Analysis of Jürgen Moltmann's Doctrine of Creation*, PhD thesis (Manchester University, Manchester), pp. 286–322. (Also published by Edwin Mellen Press, forthcoming)
48. The Eastern Orthodox view is that it is the responsibility of humans to act as the link between God and creation, bringing all of creation into communion with God. See J. D. Zizioulas, 'Preserving God's Creation', *Kings Theological Review*, vol. 13, issue 1 (1990), pp. 1–5.
49. K Rahner, *Theological Investigation*, vol. 1, trans. C. Ernst (Darton, Longman and Todd, 1965), pp. 302–17.
50. This concern has been raised forcefully by a number of different writers. See, for example, S. McDonagh, *To Care for the Earth* (Geoffrey Chapman, London, 1986); D. Dorr, *The Social Justice Agenda* (Gill and Macmillan, Dublin, 1991).
51. A. Linzey, *op. cit.*, p. 143.
52. Aristotle, *Politics 1/VIII*, trans. E. A. Sinclair (Penguin Books, London, 1985), p. 79.
53. T. Aquinas, *Summa Contra Gentiles, III*, chapter CXII, reprinted in *Animal Rights and Human Obligations*, 2nd edn, T. Regan and P. Singer (eds) (Prentice Hall, New York, 1989), p. 56.
54. A. Agius, *God's Animals* (Catholic Study Circle for Animal Welfare, London, 1970), pp. 8, 17–18.
55. See, for example, J. Moltmann, 'The Cosmic Community: A New Ecological Concept of Reality in Science and Religion', *Ching Feng*, vol. 29 (1986), pp. 93–105. Note also that for McFague sin is a refusal of humanity to live conscious of our relationship with other human beings, animals and 'nature', a 'refusal to accept our place', S. McFague, *The Body of God* (SCM Press, London, 1993), pp. 112–29.
56. A. Linzey, *op. cit.*, p. 138.
57. The Ecumenical Patriarchate, *Orthodoxy and the Ecological Crisis* (World Wide Fund for Nature, International, Gland, 1990), p. 11.
58. P. Sherrard, *Human Image: World Image* (Golgonoosa Press, Ipswich, 1992).

59. B. Szerszynski, 'The Metaphysics of Environmental Concern – A Critique of Ecotheological Antidualism', *Studies in Christian Ethics*, vol. 6, issue 2 (1993), pp. 67–78.
60. M. Fox, *Original Blessing: A Primer in Creation Spirituality* (Bear and Co., Mexico, 1983), p. 49.
61. R. A. Rappaport, *Ecology, Meaning and Religion* (North Atlantic Books, Berkeley, 1979).
62. J. Milbank, 'Out of the Greenhouse', *New Blackfriars*, vol. 74 (1993), pp. 4–14.
63. C. Deane-Drummond, 'Recalling the Dream: Celtic Spirituality and Ecological Consciousness', *Theology in Green*, issue 7 (1993), pp. 32–8.
64. H. Küng, *Global Responsibility* (SCM Press, London, 1990).
65. R. C. Lewontin, *The Doctrine of DNA: Biology as Ideology* (Penguin, London, 1993).

5

Environment, development and human justice[1]

> There was indeed a time when Environment and Development were not on speaking terms . . . Those who loved the environment regarded those who championed progress and 'development' as the enemy, and those who wanted to develop the poorest peoples and nations felt that much of the 'environment' would have to be sacrificed in this worthwhile endeavour.[2]

It is only comparatively recently that there has been an attempt to examine issues in environment and development together, rather than separately. The shift in attitude by the Greens came as a result of the realization that poor nations are forced to destroy their environment in order to survive and by the developers as a result of a new awareness that policies which destroy the very basis for change are invalid. The broad agreement in the notion that both environment and development are interconnected applies regardless of the particular models of development or environment.[3]

The Bruntland Report, also known as *Our Common Future*, published in 1987 by the World Commission on Environment and Development, was one of the first attempts on an international scale to bring the two issues together. However, the report seems to evade how to reconcile growth with ecological limits.[4] The culpability of lending agencies is not addressed in the report, which seems biased towards analysis of development simply in terms of economic growth. Since then, the Rio Summit of the United Nations Commission on Environment and Development (UNCED) in June 1992 attracted over one hundred heads of state with 178 governments represented. This much publicized event was, in effect, a programme for sustainable development on a global scale. The specification of the action needed to reconcile development with environmental concerns is enshrined in Agenda 21. This places a strong emphasis on people and their communities; nonetheless there are serious flaws in the document, not least that it seems to be biased in favour of Northern perspectives.[5] In this chapter I will be examining the following questions: How has liberation theology responded to these changes? How does this ideal of sustainable development fit in with other models of development? What is the real agenda behind the idea of sustainability? In what ways can theology, including feminist theology, critique current trends? What issues are raised by the notion of environmental justice? How can we move on after Rio in practical ways?

Liberation theology in a development context

Liberation theology identified with a particular model of development which arose at a time when the global and political implications of environmental issues were largely ignored. It is, therefore, hardly surprising that environmental concerns are rarely, if ever, mentioned in classical texts. The purpose of this section is to explore both the challenge of the environment to liberation theology and its possible contribution to an inclusive environmental theology. Some models of development are more likely to be compatible with particular environmental philosophies.[6] What contribution, if any, can liberation theology make to the current debate?

In the post-war era the conventional wisdom of the so-called 'modernization' theory held that in order to achieve development, so-called 'underdeveloped' nations must adopt a profit incentive and find ways and means for economic productivity.[7] More 'advanced' countries provided the missing components in order to 'boost' the fledgling economy. The implication is that economic growth creates conditions for democracy, so that economic and democratic stability are part of the same package.

The alternative view, known as the 'dependency' theory, gradually emerged and set out to challenge the modernization thesis. Paul Prebisch, as director of the United Nations Commission on Latin America, laid the groundwork for the subsequent development of dependency theory by pointing out the negative trade effects of contact between 'underdeveloped' and 'advanced' countries.[8] Dependency theory drew on some Marxist principles in that it attacked the capitalist system. Nonetheless, it was a modified version of Marxism since Marx believed that the economic and technical components were necessarily provided by the 'advanced' countries.[9] In Marxist philosophy socialism could only be successful after capitalist modes of production had produced the requisite wealth.

André Gunder Frank, writing in the late 1960s, was one of the pioneers of dependency theory.[10] He believed that underdeveloped nations stayed that way in order to support further growth of advanced capitalist countries. Frank called for a revolutionary break from capitalism. Latin America would remain 'stagnant' because any accumulated capital was appropriated by foreign monopolies or domestic elites. The 'development of underdevelopment' were two sides of one coin. Fernando Henrique Cardoso's version of dependency theory was more modest.[11] While he agreed that Latin America could only be understood by reference to its dependency on advanced capitalist nations, he focused primarily on internal social factors as opposed to external agents. His ideas about dependency which challenged the concentration of power in elite minority groups raised political questions. Both Frank and Cardoso have been criticized for oversimplification, though Cardoso at least attempts to avoid general theorization and concentrates on specific situations of dependency.

Many Latin American liberation theologians have drawn on dependency theory in an explicit way as a basis for their theological reflection. Gustavo Gutiérrez prefers the term 'liberation', rather than 'development', arguing that both processes are correlated:

> The term liberation avoids the pejorative connotations which burden the term development. At the same time, it is the logical expression of the most profound possibilities contained in the process known as development. In addition, the term development somewhat obscures the theological problems raised by this process. By contrast, the expression liberation leads us easily to the biblical sources which inspire the presence and action of man in history.[12]

Gutiérrez draws most heavily on Cardoso for his analysis of dependency theory. In his more recent work *The Power of the Poor in History*, he states that 'external dependency and internal domination are marks of the social structures of Latin America'.[13] He insists, further, that dependency theory needs to situate itself in the framework of the class struggle, both at a national and international level. He concludes that revolutionary socialism is the only way to break the bonds of an unjust society. Yet the model of socialism he proposes is not one that takes its cue from the 'concrete' forms that are present in the world, but sets out to question all dominant ideologies. The political issues raised by internal domination by elite groups are a focus for his theological reflection. It is easy to see how, given this attention to internal political questions, environmental issues do not appear on the agenda. He believes that in spite of its shortcomings dependency theory 'has by and large been a boon', since in the 1960s it 'helped the popular class to reject the politics of compromise and conformism during that decade'.[14] Furthermore, he agrees that the fundamental problem is not so much conflict between the rich centre nations and the poor periphery, but between social classes.[15]

Leonardo Boff also portrays Latin America as dependent. Although he uses Frank's analysis, he cautions against giving dependency theory any status higher than part of an ongoing investigation. A constant theme is that both Marxist analysis and social theory are useful instruments, rather than the final word. For: 'The oppressed are more than what the social analysts – economists, sociologists, anthropologists – can tell us about them. We need to listen to the oppressed themselves.'[16] Some of Boff's remarks are related to the fact that liberation theology has been severely attacked for its Marxist leanings by church authorities.[17] Furthermore, the economist Peter Moll argues that liberation theologians still take dependency theory for granted in ignorance of the economic data.[18] He is somewhat scathing in his remark that 'they were intellectually ill-prepared to assess the merit of competing economic theories of development'.[19] He also criticizes their reliance on a form of dependency theory which, according to him, 'not only obscured their understanding of economic justice, but also diverted attention away from one of the most important sources of economic weal or woe, namely national policy'.[20] Moll, in his turn, has been criticized for failing to appreciate the contribution of liberation theologians to the dependency debate. More important, perhaps, Moll seems to screen out other voices by his insistence that all need to become experts in economics.[21] Perhaps it is fair to say that while dependency theory raises a number of broad questions about social, political and economic structures, it fails to point to concrete solutions. Liberation theologians have used dependency theories as a starting point for their theological questions and subsequent resolution. However, the underlying social and political questions remain unresolved.

Alternative 'green' models of development

Red and green strategies for development

While the details of dependency theory have come under fire, the search for alternative models of development has led to a much greater recognition of the importance of locating development in the heart of the local community. The idea of radical alternative development was pioneered by the Dag Hammarksjöld Foundation in the late 1970s.[22] Emerging theories of development have to take into account the growing recognition of the political and global importance of environmental issues. As I indicated in the introduction to this chapter, the recognition that development and environment represented overlapping concerns is a relatively recent phenomenon. Following this, as Bill Adams points out:

> It is indeed only rarely and recently that environment and development have been linked theoretically with any kind of success. This has now been done by arguing the need to set environmental and resource use in a social and political as well as an economic context.[23]

It is of interest that liberation theology, with its stress on social and political issues within the economics of development, seems to be well placed to make a contribution to the current debate. It is not enough to consider areas of development and environment side by side: points of intersection have to be examined. The complex and multidisciplinary nature of this task seems daunting. Nonetheless, in practice these links are clear enough: development and environmental degradation existing in parallel as means of oppression of the poor.

How far can we link 'red' and 'green' thinking at a theoretical level? There are, at first sight, areas of conflict between red and green approaches which have led to a certain aloof treatment of environmentalists by socialists and vice versa. For example, while the oppressive system for 'reds' is the capitalist economy, that for 'greens' is the technological culture. The vision of the green movement is a communal, rather than a socialist society; it is achieved by revolution on a mass scale by the reds and a small scale by the greens. Above all, green strategy is decentralized, informal and in small neighbourhood communities, while red strategy is centralized, abstract and in large industrial cities, focused on the working class rather than marginalized people.[24]

The red critique of environmentalism as such is that it tends to take an essentially pessimistic view of the possibility of social change. It is also ahistorical in orientation and fails to recognize the social dimensions of environmental problems. This leads to an oppressive ecofascism where groups such as The Environment Fund were founded in the 1970s in the USA, dedicated to population control. Such an environmentalist vision of the future predicted non-development of the poorer communities of the South, leading to the pessimistic suggestion that 'there is no prospect of change in the Third World that would substantially improve the lives of more than a few people'.[25]

However, while it is fair to say that some models of environmentalism conflict with some models of socialism, it seems to me that there are ways that each can serve to critique the other in a fruitful dialogue. A model of green development proposed by Friberg and Hettne is to begin with communities, rather than expansionist industrialization. The values of a community need to be reflected in its development, so that self-reliance fosters its sense of dignity and worth.[26] Furthermore, this is in the context of social justice and 'ecological balance', that is awareness of local and global limits. Here social justice implies not just a simple redistribution of wealth, but access to wealth, education and decision making. Hence, a radically different green form of development would not only be endogenous, but also ecologically sound, that is 'utilising rationally the resources of the biosphere in full awareness of the potential of local eco-systems as well as the global and local limits imposed on the present and future generations'.[27] While this model is far from a naive critique of environmentalism as ecofasicst, it has failed to incorporate the political aspects of 'red' thinking, in other words it still tends to remain parochial, setting the agenda to local contexts without consideration of the structural barriers to change. How far does the concept of endogenous development really challenge capitalism or socialism?

Bahro sets out in a more deliberate way to link 'red' and 'green' thinking in his *Socialism and Survival* and *From Red to Green*.[28] In contrast to Friberg and Hettne's thesis, his 'red-green' ideas offer a rigorous double critique of both capitalism and current socialism. The focus of his attack is industrialization as such with associated increase in material consumption and production. Hence developmental models which try to increase trade and industrialization in the Third World are like 'a tunnel without an exit'.[29] For industrialized Western nations an escape was possible through exploitation of the periphery, but for the nations of the South no escape is possible.

What is sustainable development?

The idea of 'sustainable development' has become an overarching concept which recognizes the links between environment and development. Yet the term itself is so vague that it can be grafted on to a number of different ideologies, including those from 'red', 'green' and 'blue' perspectives.[30] In a broad political context it is usually the sustainability of human society that is being referred to. Sustainable use is part of the general concept and refers to the need to ensure the use of resources to allow the sustainability of human society. In a more specific sense sustainability 'refers to components contributing to the sustainability of society, that is: population, consumption, resource use and pollution'.[31] The complex nature of each of these components means that it is far from clear what constitute sustainable policies. For example, resources are distinguished by their renewable and non-renewable nature, whether they are living or not and whether they are 'natural' or 'man-made'. Protagonists of sustainability can allow for the substitution of one resource for another, without taking into account the global environmental effects.[32] The main ethical thrust behind the idea of both sustainability and sustainable development seems to be the obligation to future generations, which is highly contentious.[33] However, if the notion of intragenerational justice is included then sustainability can be linked with present and future development.

Some environmental ethicists have rejected the idea of sustainable development altogether as 'irredeemably anthropocentric'.[34] It seems to me that this is only true if sustainability is used in a narrow sense to describe economic factors leading to reduced environmental risk. These risks have to be weighed up in order to maintain a privileged position in the market. In practice this means that restraint is based on the legality or otherwise of the action in a particular nation. Bill Adams argues that the concept of sustainability as such cannot escape elements of control, both of nature and of people.[35] He believes that sustainable development depends on a vision of managerialism which is drawn from ecology, and in turn from systems theory. This theory, which stems from the sciences of engineering and thermodynamics, fosters an implicit notion of control: that development is still essentially a managerial project.

An alternative, broader view of sustainability is one which is not just about economics but is an ethical ideal of life on earth. This approach recognizes that environmental value is much wider than that measured by economic means. It includes a holistic approach to the natural world which stresses the inseparability of human beings and their environment. The problem with this approach is that the needs of the individual in development can be overlooked.

The greening of economics

There are two major schools of thought suggesting the ways in which we could 'green' economics. One way is by an essentially revisionist or neo-classical approach. This leaves the market economy intact and is naturally more favoured by governments, intergovernmental institutions and multi-national companies. According to this school of thought, taxes or other financial incentives such as pollution permits can be used to encourage more environmentally friendly behaviour. The introduction of a 'green' tax on products and/or activities that pollute, degrade or damage the environment, ensures that environmental effects are taken into account in the planning process. I will return to the issue of environmental impact assessment in the section below. The advantage of this system is that it is targeted directly at those responsible for damage, rather than raising revenue by a general tax increase. Brown *et al.* believe that such green taxes should not supplant legislation for the environment, such as towards the protection of endangered species, but serve to complement it.[36] The advantage of this model is that it is easier to achieve though existing structures of government and it does at least go some way towards alleviating environmental damage.

A second major school of 'green' economics is that which calls for more radical reconstruction of the whole economic system. Herman Daly is one of the leading proponents of this view.[37] He argues that simple tinkering with the existing economic system does not go nearly far enough. What is needed is a radical redesign of economic parameters so that the aim is for a steady state economic system, rather than the 'growth mania' which has afflicted the world economy. He suggests that we should aim to work so that our activities achieve equilibrium with the environment, rather than aiming for growth regardless of the environmental consequences. He argues that the market, by itself, has no way of introducing an environmental criterion, its main goal is towards growth. He suggests that micro freedom and variability can be combined with macro stability and control by fixing the total resource 'collectively on the basis of ecological criteria of sustainability and ethical criteria of stewardship. This approach aims to avoid both the Scylla of centralized planning and the Charybdis of the tragedy of the commons.'[38]

Daly suggests that there need to be some restrictions on food supply and resultant pollution. According to his scheme the total limits are set by the government agency through depletion quota permits. Once these have been purchased, private companies operate within these total limits. He also suggests a birth quota, though it seems to me that this is outside the boundaries of acceptability in a democratic society. One major difficulty with this system is that it assumes that all countries are in a position to reach steady-state growth simultaneously. He acknowledges that the Third World needs to be given permission to grow up to certain limits, but the question is: Who is to set these limits? A further difficulty is the assumption that it is possible to control the economy in this way: the model seems to be close to the one which Adams rejects, namely a technocratic approach relying on the thermodynamic and control systems theory that I mentioned briefly above. It is of interest that the *Reader* in sustainable development finished with a reflection on alternative economics, as if economics alone is the key to the resolution of the problems facing sustainability. However, I wonder if this supposed radical view is still too lopsided? Is it too mechanistic and arrogant in supposing that we can successfully manage the earth for the benefit of all creatures, including the human species?

Towards 'holistic' development?

In view of the ambiguity of both the terms sustainability and development, I prefer the term 'holistic development' to imply a model of development which is organic rather than mechanistic. However, I do not intend to imply that I am endorsing the 'deep green' philosophy of Naess and others.[39] In speaking of the ethics of environment and development there is a tendency to set up radical alternatives of either a mechanistic and Cartesian world view or an ecological and holistic world view.[40] The former is characterized by splits between fact and value, ethics and life, subject and object. 'Nature' is seen as discrete, material resources with instrumental value. This view seemingly leads to a centralization of power, competitive attitudes and undifferentiated economic growth. The latter is characterized by a close relationship between fact and value, ethics and life, subject and object. 'Nature' is made up of interrelated wholes which are given intrinsic as well as systemic value.[41] It leads to decentralization of power, a multidisciplinary approach, cooperative thinking and steady-state economy. When the alternatives are posited in this way it is hard to visualize any alternative other than radical replacement or revolution. Furthermore, there is a tendency in 'deep green' philosophy to let 'sustainable development' become 'ecological sustainability', detached from the issues of human justice.

I hope to show in the section which follows how a delicate balanced link between environmental sensitivity, human justice and development can become fostered through a revised form of liberation theology. The language of revolution identifies with the early writing of liberation theologians. How far has this language become modified in view of the practical realities of implementing change? It seems to me that liberation theology may have an important contribution to make in discerning practical ways forward to a more holistic development.

Emerging links between liberation theology, environment and development

The focus on endogenous development at the heart of the local community, which is characteristic of some models of alternative development, coheres well with Christian base communities in Latin America fostered by the ideals of liberation theology. Robertson challenges the claim that Christian base communities are vehicles for liberation theology and suggests that, in practice, their purpose can be more ambiguous.[42] Thierry Verhelst, by contrast, has argued that liberation theology has acted as a powerful impetus for resistance to cultural alienation.[43] Nonetheless, he cautions against a simple adoption of liberation theology by other cultures. He comments:

> In the East the term liberation has different connotations than in Europe or Latin America. Asia is the cradle of all the major written religions. One would hardly expect its conception of liberation to be secular or basically socio-economic. For both Hinduism and Buddhism, liberation is achieved through a primarily spiritual, inner experience. . . . The message of Eastern spirituality is the following: it is not only exploitation, domination and material poverty that ought to become objects of the struggle for liberation. There is another poverty, at least as serious: that is engendered by self-interest and egocentrism. In the East this poverty is called *Maya*, illusion, and becoming aware of it represents the starting point of its spiritual journey.[44]

The theme of liberation is especially important in Latin American liberation theology, while that from an Asian or African origin tends to stress inculturation. The radical thrust of Latin American liberation theology has gradually moderated under the influence of the concept of enculturation. To the Eastern mind the preconceptions of liberation theology are still Western, even though liberation theology seems more equipped to answer the needs of the Third World.[45] An Asian theology and spirituality of liberation 'finds its inspiration above all in the spiritual asceticism of the individual, for the oriental tradition of renunciation cannot be ignored'.[46] In Asia the totality of human experience – personal and communal, mystical and social – replaces the social praxis of Latin American liberation theology. It seems to me that this emphasis on the totality of experience is more amenable to a theology for holistic development. Even though there is truth in the suggestion that theologies and spiritualities have to emerge from local culture, it is possible to learn from and be challenged by alternative perspectives. If total enculturation took place to the exclusion of external influences, I doubt whether there could be any sense of global community. In other words, the practical vision of what is possible at a local level can be modified by global considerations. The ecological crisis would seem to demand such global interests.

However, Verhelst's designation of East or West liberation theologies does not take into account the contribution of indigenous theologies, including native American theologies. In a recent book, *EcoTheology: Voices from South and North*, environmental questions are placed firmly in the context of human ecology and development.[47] The role of indigenous cultures is stressed, but there is a tendency for overgeneralization, such as 'the underlying cause is essentially the replacement or domination of the "culture of life" of the indigenous, non-industrialised peoples by a "culture of death" characteristic of the rich and powerful in the industrialising and industrialised countries'.[48] The theologies from indigenous cultures, or the Fourth World, wish to stress their distinctive contribution in terms of a revelation of God in place, taking their cues from Old Testament themes. They believe that the Third World liberation theologians still take their cues from Western culture. Indigenous cultures, in contrast, take their cues from identification with the land. Rob Cooper expresses the wish thus:

> Our hope is that in coming so late into a world mad with materialism, our identification of ourselves as literally 'people of the land' and our harmony with our environment will reflect the way the world could be. There remains in Maori culture, in the way we live, that sense of unity in creation . . . Naming and knowing our world distinguishes us from the brutes. It also joins us together in responsibility for it – and them.[49]

While, as I noted above, blanket condemnation of Western culture seems unjustified, the Western emphasis on a mechanistic approach to creation has much to learn from indigenous cultures such as these. Above all they reintroduce the idea of a theology of place.[50] The liberation theologies of Latin America have failed to recognize the close identification of indigenous people with their land, in assuming that it is economic deprivation that is at the heart of injustice. In other words, their sense of being persons who are culturally distinct comes from this identification with the land. A Marxist solution is not necessarily the answer that is desired:

> To put the means of production into the hands of the poor eventually makes the poor exploiters of indigenous peoples and their natural resources. Finally it seriously risks violating the very spiritual values that hold an indigenous group together as a people.[51]

It seems to me that the danger that classical liberation theology can become yet another subtle means of oppression of some groups is real enough. Nonetheless, there is also the danger that we can adopt an overly romantic view that assumes that a simple global return to more animist spiritualities provides a solution to the problem. Instead we need to engage in dialogue with other cultures, while respecting their differences.

The above is illustrated further by attempts to adopt developmental policies of the so-called 'green revolution' which are out of step with indigenous practices. Third World women, for example, use grasses that grow alongside the borders of fields to make baskets and mats:

> When development experts decide that these grasses have no market value and plan to carry out programmes to kill them with herbicide, they rely on an arrogant perception of their relationship to nature. Far from defending the world's poor, the green revolution sought to defeat communism by destroying the world's peasant class. It did so by dividing the peasants economically, politically and spiritually from their sense of place. Conversely, one could understand much about peasant resistance movements by considering them to be localised defences of the connection between person and place. It is no accident that liberation theology in Latin America or Dalit theology in India are perspectives that express the particular conditions of specific peoples. They reject the universalising tendencies of traditional theology.[52]

Liberation theology, by looking to rather different models of liberation in Asia and the Fourth World, can begin to become more inclusive of environmental interests that are still bound up with their own interests. Leonardo Boff argues for a social environmental ethic, one which restricts the behaviour of human beings towards each other, as well as the behaviour of humanity towards the natural environment. He argues that social injustice against peoples and environmental injustice against the earth affect people directly and indirectly. I welcome his suggestion that we need to avoid the extremes of anthropocentrism and naturalism, which

I discussed in some detail in the previous chapter. He is strong in his claim that the earth has 'rights', which he justifies on the basis of longevity; in other words because the earth has existed millions of years before humanity, it has a right to exist in 'well being and equilibrium'. I am less convinced that this presents a valid case for right treatment of the earth; I prefer a more theocentric approach to environmental ethics. He admits that liberation theology, as well as other theologies, has emerged without adequate recognition of the environmental context. A more holistic vision is required where we admit to ecological sin and allow the planet to become 'a great sacrament of God, the temple of the Spirit, the place of creative responsibility for human beings, a dwelling place for all beings created in love'.[53]

Another significant book in this context is a recent issue of *Concilium*, edited by Leonardo Boff and Virgil Elizondo and entitled *Ecology and Poverty*.[54] The book knits together issues of poverty and ecology and stresses the sacred earth traditions in indigenous cultures. Boff's contribution is particularly relevant for the present discussion where he asks whether liberation theology and ecology are alternatives, confrontational or complementary.[55] He seems to be coming close to the deep green philosophy of Naess when he suggests that: 'Liberation theology should adopt the new cosmology of ecological discourse, the vision that sees the earth as a living superorganism linked to the entire universe.'[56] While I identify with his suggestion that liberation theology and ecological discourse can mutually complement one another and act as a bridge between North and South, I am unsure of his seeming ready identification with radical elements of the ecological movement. It is, however, natural that he identifies with radical ecology as this offers the strongest challenge to capitalism and the status quo and is more in tune with the revolutionary message of classical liberation theology.

In practice, however, contemporary liberation theologies have had to argue for reform, rather than revolution. The Christian base communities have not always proved to be the centres for revolutionary change implied by some liberation theology. Hewitt argues that the primary benefit of base communities has been to endear a sense of citizenship.[57] There is also a corresponding shift in the writing of liberation theology from Marxist to democratic ideals, from conflict to negotiation, from class struggle to solidarity with the poor.[58] In pressing for change in environmental/developmental policies a more moderate view which argues for reform, rather than radical change seems to me to be the more realistic option. The radical vision can remain, but an intermediate stage of reform may have to be tolerated for some time to come.

The critique of feminist theology in relation to environment and development

The theme of liberation can also apply to the relationship between human beings and the natural environment. First, there is liberation from an attitude which views science as an instrument of power over 'nature'. As Kothari points out:

> The presumption that the role of science and technology was to develop nature in the service of humankind has turned out to be an illusion. It was based on a view of science itself as an instrument of human power over nature, other men and women, other forms of life and all the qualities of being that constitute the cosmic order. This must give place to the original purpose of science, namely seeking to understand the mysteries of nature with a deep sense of mystery and wonder.[59]

The link between oppressive development policies and the oppression of women has been the focus of concern for feminist writers. Modern scientific world views are scrutinized by Shiva, who views them as 'a masculine and patriarchal project which necessarily entailed the subjugation of both nature and women'.[60] Women are identified as those who are pioneers in the struggle for the protection of the earth. Feminist theologian Rosemary Radford Ruether argues that: 'Rebuilding human society for a sustainable earth will require far more than a plethora of technological "fixes" within the present paradigm of relations of domination.'[61] She argues instead for 'biophilic mutuality', which takes into consideration transformed political and cultural consciousness. Her main theme is the empowerment of women and a transformation of men from the 'illusion of autonomous individualism', but not in isolation from a larger transformation of women towards a 'holistic' culture.[62] She believes that this transformation is possible through building 'base communities' of celebration and resistance. It is interesting that she uses the language of liberation theology to express her ideals. Such communities learn how to 'be', rather than how to 'strive', ultimately thinking and acting globally as well. While I can see that her vision of community life offers a new way of thinking about lifestyles, there is a danger that such communities will fail to look outward to the political and economic realm and become instead ghettos of warmth and support. How far do they really challenge the socio-industrial complex? How far does this really engage with complex issues in the North and South? To some extent this issue is addressed in a recent book edited by Ruether entitled *Women Healing Earth* and written by a number of different feminist writers from Latin America, Asia and Africa.[63] The most significant theme of this book is the emphasis on the cultural context of these women and their reinterpretation of sustainability as an issue for transforming women's roles. Nonetheless, there seems to be an underlying assumption of the value of women's identification with the earth, rather than distinction from it.

The celebration of the supposed link between women and nature brings certain difficulties. How far, for example, does it reinforce the stereotypical view of woman as 'earthy' and 'maternal', ending up with a narcissistic loss of distinction between self and the other? There is, furthermore, a tendency to universalize the attitude of women (and men), thus ignoring factors such as race and culture. An alternative approach is to allow all cultural expressions to coexist and speak from within that context, in a way which seems to lead to the promotion of relativism. The swing towards this second alternative is clear in *Women Healing Earth*, mentioned above. While the rootedness of these studies is commendable, I am left wondering how these ideas could be used to challenge Western cultural values and society. In as much as they speak out of a context, they remain in that context, liberating the oppressed, rather than the oppressors. While there has been much talk of restructuring of science from a feminist perspective, the reality of the shape of this science remains vague.[64] The liberation of women and nature from oppression cannot be resolved by restructuring science; rather this seems to me to be a political issue, to which I will return again below. The goal of industrialization and technology is one which depends on consumerist demand for more goods. It is not, then, so much the transformation of science, but the reorientation of its aims and goals which seems to be important. I am far from convinced that a simple rejection of 'dualism' of humanity and 'nature' is the key issue here. Indeed the loss of distinctiveness seems to lead to a passivity and apathetic approach to the natural world. Empowerment implies an assertiveness that can affirm difference, but refuses to surrender to domination by the other.

Irene Dankelman and Joan Davidson take up the issue of women's empowerment as a development issue.[65] The continuing population growth in the South increases the struggle for survival, which in turn means higher mortality rates for infants and consequent pressure for larger families. Iniquitous practices include distribution of unsafe and inappropriate contraceptives. While an integrated approach to family planning and environmental action is one way of approaching the problem, it seems to me that it still leaves the underlying causes of poverty intact. A particular concern for the Roman Catholic Church is how far particular birth control policies really take into account the global ecological issues and the strain that overpopulation is making on the earth. Statements from the Roman Catholic Church which have implied that the population question is not a serious threat have alienated those who are well aware of the potential unsustainability of the current global population explosion.[66] The statement by the Pope in the document *Humanae Vitae* that 'each and every marriage act must remain open to the transmission of life' reaffirmed the traditional teaching of the church on contraception. This document has led to a huge controversy among priests and lay members. Even though it carries the personal authority of the Pope, many Catholics believe that it is not possible to adhere to the teaching. This is especially true in the context of the burden of overpopulation on the earth's resources. The way forward is through empowerment of individuals in their communities, rather than additional oppressive measures.

Jürgen Moltmann has also addressed the issue of empowerment in his belief that modern industrialized nations have led to humanity becoming trapped in a web of their own making.[67] Liberation becomes an all embracing concept which includes economic justice in the face of human exploitation, human dignity in the face of political oppression, human solidarity in the face of alienation and division, peace with nature in the face of industrial destruction and hope in the face of apathy towards the whole.[68] Moltmann recognizes interconnections between different forms of oppression and so, by implication, the need for cooperation between different forms of liberation. A similar point has been made recently by Boff and Elizondo in their identification of the 'Cry of the Earth' with the 'Cry of the Poor'.[69]

The idea of liberation has been taken up by Gerald Kruijer, writing from a secular sociological perspective.[70] He believes that the social sciences have to liberate themselves from the need for abstract concepts and the quest for unbiased objectivity. A liberation science is science in the service of the liberation movement. He argues that data collected in the past have been biased against the poor. Moreover, everything that exists, 'including non-living matter', becomes part of the solidarity group.[71] The results of applied research done in this way have to be fed into policy formation. He argues that there is an intermediate stage between oppression and democratic socialism, namely undemocratic state socialism. He also offers some suggestions for socialist strategies and how to move from state socialism to democratic socialism. While many of Kruijer's ideas are controversial, he does at least offer some suggestions as to how to begin to implement realistic change, which seems to be lacking in liberation and feminist theology. Taylor makes a similar point when he suggests that the real debate in development policy is not so much what is good, but what will work. An ethical critique can examine motives and consequences of policies, as well as offering some directives for future policy formation. In particular it will ask whose interests decisions serve and what the hidden agenda behind policy decisions is.[72]

The problem of environmental justice

It is ironical, perhaps, that environmental issues can become a source of conflict arising out of a perception of injustice. Conflict over 'ownership' of environmental resources, especially those shared between nations, can become the catalyst for militarist action and political instability. The UNCED Summit in Rio was, in one sense, recognition by governments and nations that international environmental obligations are a crucial issue in international affairs. Environmental treaties bring with them environmental obligations as well as possible threats to security through the breakdown in these treaty agreements. Legal systems have been put in place for over a century as resources diminish.[73] However, today controversy exists over issues such as transboundary air pollution or diversion of water supplies from rivers crossing international boundaries. While earlier international law dealt with localized issues, such as the protection of endangered species, the modern context is one which recognizes the integration of environmental issues into development activities. Sands suggests that: 'The number of environmental disputes undoubtedly will rise as the international community seeks to reconcile the conflicting demands of economic growth and environmental protection.'[74]

One source of international tension is that in less developed countries there is a strong belief that the resources which were plundered by the colonial powers should be restored. This does not relate simply to North–South relationships; it can also apply to the newly independent states of Eastern Europe, for example. One case study is that of the island of Nauru, which prior to its independence in 1968, was subject to phosphate mining on such an unprecedented scale that one third of the land was rendered uninhabitable.[75] The island is demanding proceeds of the phosphate sales, in order to begin to try and repair the damage, through the international court of justice. Nauru is drawing on the general principle of international law that a state responsible for the administration of a territory must not cause irreparable damage to the legal interests of a successor state. The dispute illustrates the serious consequences of irresponsible behaviour and could be applied to similar disputes among other states and nations. For example, is the UK liable for the damage caused to Australia as a result of nuclear tests prior to independence? In this case is the UK culpable, even though it acted largely in ignorance of possible consequences?

Another issue of environmental justice is the excessive level of pollution tolerated in the Third World at a national legal level. This has left a loophole for those unscrupulous towards the needs of the poor. The examples of inappropriate 'development' are legion. One is the ever increasing use of pesticides in the Third World. While the Third World uses only 15 per cent of pesticides at a global level, 50 per cent of cases of poisoning occur in these countries.[76] In northeast Sri Lanka, for example, 938 deaths from pesticides in 1977 exceeded the total mortality from malaria, polio, tetanus, diphtheria and whooping cough. The inappropriate labelling of pesticides, ignorance of their correct use and lack of appropriate equipment to apply the chemicals, all contribute to this problem. Moreover, chemicals considered dangerous for Northern nations, such as DDT, are still used. The accumulated effect of industrial plants which lack appropriate 'clean up' facilities for their waste and increased use of pesticides contribute to the rising levels of pollution damage in Third World countries. There is a lack of any accurate data on the actual damage caused by pollution, so that the overall environmental impact is unknown and unacknowledged.[77]

A related issue is that environmental protection against pollution is biased in favour of some communities over others. Minority communities are disproportionately subjected to a higher level of environmental risk.[78] Even the assurance of equal protection does not take into account the fact that disadvantaged communities are more vulnerable to pollution because of poor health.[79] The environmental justice movement presses for changes in environmental policy in order to take this into account. In the past concern with environmental issues has been thought of as a middle-class luxury which pales into insignificance in the face of questions of survival amongst the world's poor. Yet it is clear that questions of survival and questions about the environment are intermeshed to the extent that the very survival of communities is dependent on taking into account environmental issues.

The ban by some states on imports of products that are considered to be environmentally damaging raises additional complex issues of environmental justice. While on one hand this leads a nation to rethink their current practice, on the other hand severe economic consequences fall on those who are ill-equipped to suffer such losses. The USA, for example, banned the import from Mexico of tuna which had been caught with fishing nets that led to a high incidence of dolphin mortality.[80] Mexico asked the General Agreement on Tariffs and Trade (GATT) to examine whether such a ban ran counter to GATT rules on free trade. Eventually the ban was lifted. The case illustrates clearly how environmental ideals can become a source of conflict between nations. There may be more appropriate ways of encouraging poorer nations to adopt environmentally friendly practices than a simple ban on their products. Above all, a form of subsidy for those who do use environmentally sound policies seems a more positive approach. The tendency to blame the poorer nations is also implicit in this case. As I showed above, the Northern nations are implicated in environmental damage through pesticide use and industrial investment in a way which makes their ban on some products seem shallow and hypocritical.

Sociological research has shown that in the developing world government conservation schemes tend to ignore the needs of the inhabitants and in some cases deprived them of their means of subsistence.[81] Non-governmental organizations fared slightly better, though very often they were 'deeply involved in conservation projects that were very similar to those of the government'.[82] If conservation is conducted in ignorance of the social consequences it is ultimately destructive as it works against the basic need for human justice.

From a theological perspective it is clear that justice issues have to be addressed within the human community as well as from a broader environmental perspective. A *metanoia* is needed at all levels of development. This metanoia includes a positive search for peaceful means of resolving conflict. As John Zizioulas has pointed out, we tend to believe that a resolution of the problem is to be found purely in ethics and/or legal restrictions on behaviour.[83] It seems to me that ethics and laws emerging from this give us useful practical outlines of the problem. They serve to clarify and delineate responsibility. However, the change of heart required to reach to the root of the crisis is as much a spiritual as a secular obligation. The liberation of humanity and the natural environment presupposes the liberation of human beings from self-centred behaviour. In other words, the political agenda of liberation theology is not sufficient on its own to generate the will to change.

Liberation theology in its more mature phases recognizes the mystical dimension, the Other that is God through whom all creation comes into communion. Humanity relates to God by participating in an offering that brings all of creation into communion with the love of God. An environmental theology is a theology from below, but one which acknowledges the transcendence of God. It seems to me that a theological paradigm which best expresses this theological approach is that of wisdom, wisdom which springs from a reflection on divine wisdom. Indeed it is hard to see how there can be any resolution of the complex problems we face without an appropriation of wisdom. As a secular term wisdom can take us so far. However, as a theological term wisdom serves to challenge injustices and human pretensions for grandeur. Hence the wisdom of God can seem like foolishness; the crucifixion can seem anathema to those dreaming of a Utopian future. Yet the double movement of cross and resurrection reminds us that whatever befalls this planet the situation is never entirely hopeless. Christian theology can serve to keep this hope alive, but as Moltmann points out, it is a realistic hope, not one caught up in illusion. For: 'the doxological anticipation of the beauty of the kingdom and active resistance to Godless and inhuman relationships in history are related to one another and reinforce one another mutually'.[84]

Examples of inhuman relationships in history that Moltmann highlights can be extended to include inhuman relationships with our environment as well, so that just as the destruction of nature is linked with our own self-destruction, so the rebirth of humanity is linked with the future hope of the whole created order. In this way liberation becomes as much an inner liberation as an outer structural liberation, fostering an authentic existence which leads to right relationships and just political systems. If theology has anything to say it is this: that the spiritual dimension of humanity is ignored at our peril, that we have a responsibility to search from within our own theological tradition and beyond for the inner resources needed to bring about lasting change. The will for transformation comes, then, from an inner transformation, the ethics from a right relationship with God. Thus practice flows from theology in the light of current concern, rather than ethics being used as a cover to promote our own particular ideologies. It strikes me that sustainability has the potential to be misused under the umbrella of false ideologies and narrowly defined interest groups. Within a theological framework it can be channelled for the good of all, not just certain sections in our world community.

Sustainable planning policies: myth or reality?

Developments since Rio

While Rio put the issue of the close interconnection between environment and development firmly in the public consciousness, there is much debate as to how much it really achieved in practice. The treaty on climate change, for example, encouraged but did not commit nations to reduce their emission of greenhouse gases. The treaty on biodiversity is similarly non-committal, with indigenous people failing to gain a share in the profits made by transnational agrobusinesses and pharmaceutical companies exploiting resources in the Third World. The protection of the remaining rainforests was not binding. There is a strong suspicion that the pressure for compromise and the overall agenda came from the rich nations of the North, most notably the World Bank and transnational corporations.[85] More positively, it was the informal networks that were established at Rio that are likely to have the most lasting benefit. The Global Forum meeting of non-government organizations (NGOs) provided the opportunity for development and environment workers from the North and South to address the social, political, ecological and economic issues which are at the heart of developmental and ecological problems.

Other developments since Rio include the establishment of Local Agenda 21 initiatives. These initiatives have, on the whole, been more successful that those conducted by central government. Global Forum 94, held in Manchester in the summer of 1994, looked at the implications of Agenda 21 for sustainable cities. Local authorities, trade unions, business and industry NGOs and community groups were represented and joined by 'international experts'. Reid is enthusiastic in his support of this meeting, stating that 'its key achievement was to demonstrate that the four sectors that were represented could be brought together to achieve consensus and find innovative solutions'.[86] He is probably unaware that there was a parallel meeting of an 'Alternative Global Forum' that met in Moss Side Community Centre. This group felt strongly that the official Global Forum was biased towards the most successful sections of society, that the needs of the poorer communities had not really been taken into account and that the meeting encouraged rhetoric rather than action. This story illustrates how the public perception of an event can influence its appraisal. Success or otherwise depends on what are the perceived goals and aims. This local initiative seemed to suffer from the same flaws as the Earth Summit in Rio itself, namely a failure to be radical enough in encouraging real participation by those most likely to be implicated in any policy decisions.

This reinforces the idea that sustainability cannot be imposed by agencies representing national or international alliances of political and economic power, but comes from the heart of a community itself. Roddick believes that a major barrier to change is not so much the bureaucrats themselves, but the lack of public will to implement change.[87] The need to raise public awareness and understanding of the problems and issues is one of the first steps in ensuring that sustainable alternatives are put into practice. Further, participation by a local community can only be effective if it is supported by government actions. Participation as a slogan will be as ineffective as sustainability if the will to devolve power by governments is lacking. The kind of constructive support that governments could provide includes resources for education, community development initiatives, land reform and changes in planning law and regional economic policy.

Environmental impacts: planning for sustainability?

The adverse ecological impacts of development have led to an essentially technocratic approach to the issue, where the aim is to produce maximum profit within 'acceptable' environmental impacts. In practice, what is acceptable depends on the legal restraints of the nation concerned. Environmental Impact Assessment (EIA), introduced in the late 1960s, has become an important part of public policy in the industrialized world. The Scientific Committee for Problems of the Environment (SCOPE), set up in 1969, worked on the basis that with adequate scientific analysis environmental impacts could be monitored and controlled. Yet assessment of potential impact demands judgements on the part of the analyst that are not 'scientific' in the strict sense. Furthermore, the jargon used is pseudo-quantitative, which removes the issue from public domain. As Bill Adams points out: 'Whether cloaked in quantification or not, such procedures are essentially qualitative, and therefore highly dependent on the skills, prejudices and perceptions of the analyst.'[88] The failure here is twofold. One is the assumption that quantification of a problem will bring it under 'control' of humanity. Attempts at quantification ignore intangible effects on culture, society and long-term environmental effects which cannot be measured or predicted. The other is that by removal of the issue from public assessment the 'expert elite' now has unwarranted power to make decisions that are not necessarily in the public interest.

The difficulty of appropriate environmental impact assessment is accentuated in the Third World. The damage done to the environment is often remote from the site of production and thus considered outside the boundary of the developers. The cost of making accurate and comprehensive environmental impact assessments is commonly outside the resources available in the Third World development projects, both in terms of finance and technical skill. Bringing in outside expertise is seen to be the 'solution', but this further fosters a sense of domination of poorer nations in the 'periphery' by those in the richer 'core' communities. The projects themselves put high priority on expertise from 'hard' disciplines such as engineering, hydrology, agronomy and economics and low priority on the engagement of ecologists or others from 'soft' disciplines. The net result is that EIAs are tacked onto the development process far too late and in a cursory manner, instead of becoming an integral part of the planning procedure.[89] Political pressures, including corruption of officials, also distort adequate appraisal, especially where the state is determined to implement certain ideas for material or other political reasons.

The International Institute of Environment and Development (IIED) has conducted research into the environmental policies of aid agencies, including the World Bank, which is the largest of the multilateral donor aid agencies.[90] The research suggested that while they employ a small number of environmentally aware individuals, the sheer size and bureaucracy of the organizations leads to relatively low impact on overall policy decisions, ecologists' ideas are rejected as 'unrealistic' and an impairment to supposed 'progress'.[91] Criticisms of the World Bank's environmental performance have continued, in spite of attempts at reform. IIED's appraisal of other major aid agencies from Canada (CIDA) and the USA (USAID) focused on the lack of appropriate expertise: the agencies were dominated by economists and planners at the expense of environmentalists and/or local involvement. Overall it is hard to see how the 'greening' of development will be achieved, as Adams remarks:

> Large bureaucracies are inherently conservative, and the 'greening' of development is bizarre theoretically in the context of the established disciplines of development planning, troublesome in terms of policy, and highly inconvenient administratively. Reforming the practice of development bureaucracies has to go a great deal deeper than the superficial transformation of rhetoric and terminology.[92]

One way to begin to approach the issue is actively to promote small-scale projects, drawing on the example of Schumacher in *Small is Beautiful*.[93] Small-scale projects have the advantage of having less adverse environmental effects, are easier to incorporate into local needs and are less wasteful in terms of resources. Nonetheless, while this idea is theoretically attractive, there are still unforeseen problems of which those involved in planning need to be fully aware. For example, small-scale irrigation schemes in Kenya have proved to fail in practice, showing problems of flooding, water shortage and salinity.[94] In other words small scale alone does not mean appropriate technology; there is no magic formula to achieve sustainable development. Above all the locus of power must shift to those who are affected by the so-called 'development'. It is all too easy to impose plans from the outside, however well-meaning they may seem even in terms of environmental objectives. There are theological and ethical reasons for putting the poor at the heart of the development process.

Conclusions

All these examples highlight the need to take into account both the social and environmental consequences of particular policies. While liberation theology originally conceived as social praxis does not take into account ecological issues, it is still relevant for discussions as it raises questions about human justice as part of a holistic approach to the environment. Furthermore, the idea of liberation can become enlarged to include the liberation of human beings from their desire for domination not just of each other through oppressive regimes, but desire for domination of the natural world.

Liberation theology can provide an incentive for the radical change in attitude that is needed at all levels of society. More extreme eco-philosophers can take heed of the adjustment of the revolutionary agenda of early liberation theology to a more modest goal of social reform. This social reform can include a reappraisal of the care for the natural environment, so that development policy becomes holistic development, embedded in an ethic of environmental responsibility. Holistic development implies an inner spiritual transformation alongside external, structural changes. If theology has something to say to the issues of environment and development it is this: that it can generate a greater sense of balance in the lives of individuals, communities and governments; that we need to be liberated from a concentration on material growth through a controlling attitude to others and nature; that liberation includes both the oppressors and the oppressed, whether we are talking about the domination of minority groups, women or the natural environment.

It seems to me that the themes of liberation can become extended and apply to whatever culture and place we occupy in society. Liberation, furthermore, implies recognition of this place and working to dismantle ourselves and those in our peer group from oppressive tendencies. It is our responsibility to gain insight into how liberation can apply in our own context. Just as Asian liberation theology is different from Latin American liberation theology, what would Western liberation theology look like? I suspect that it would, to some extent, accept that we start from a culture which gives value to individual achievement. As individuals we cannot hope to make more than a modest impression on our world. Liberation theology has shown Western theologians the impact that a small group of dedicated individuals can achieve through collective action. While liberation theology for the Western mind is, as I have implied, an individual one, we can surely argue that this is not the same as individualism, which elevates the individual above community. The global justice issues and global environmental problems will not go away as long as we act just as individuals, detached from the political and community networks which have been eroded in the name of 'development' and 'progress'.

Those 'underdeveloped' nations are in fact far more 'developed' in their sense of community identity than richer Northern nations. Holistic development implies not a one-way process, but a mutual learning of each from the other, a shared wisdom between nations, while respecting the identity and difference in the other.

NOTES

1. Some ideas for this chapter are published in a condensed form under the title: 'Development and Environment: In Dialogue with Liberation Theology, in *New Blackfriars*, vol. 78, no. 916 (1997), pp. 279–89.
2. From foreword by Lloyd Timberlake in *Dictionary of Environment and Development*, A. Crump (ed.) (Earthscan Publications, London, 1991).
3. For scientific discussion see, for example, *Environment, Population and Development*, P. Sarre (ed.) (Open University, Milton Keynes, 1991).
4. D. Reid, *Sustainable Development* (Earthscan, London, 1995), p. 59.
5. *Ibid.*, pp. 192–5.
6. See D. Goulet, 'Development Ethics and Ecological Wisdom', in *Ethics of Environment and Development: Global Challenges, International Response*, J. R. Engel and J. G. Engel (eds) (Belhaven Press, London, 1990), pp. 36–49.
7. For the steps needed for this economic growth, see W. W. Rostow, *Stages of Economic Growth* (Cambridge University Press, Cambridge, 1991).
8. For a discussion, see A. F. McGovern, 'Latin America and Dependency Theory', *This World* vol. 14 (1986), pp. 104–23.
9. See Preface to *Capital* in *Marx and Engels: Basic Writings on Politics and Philosophy* L. S. Feuer (ed.), trans. S. Moore and E. Aveling (Collins/Fontana, Glasgow, 1969), pp. 174–87.
10. A. G. Frank, *Capitalism and Underdevelopment in Latin America: Historical Studies of Chile and Brazil* (Monthly Review Press, New York, 1969).
11. F. H. Cardoso, *Dependency and Development in Latin America* (University of California Press, Berkeley, 1979).
12. G. Gutiérrez, 'The Meaning of Development: Notes on a Theology of Liberation', in Sodepax Report (The Committee on Society, Development and Peace) *In Search of a Theology of Development* (World Council of Churches, Geneva, 1969), pp. 125–6. He makes a similar point in *A Theology of Liberation*, rev. edn (SCM Press, London, 1988), pp. 16–17.
13. G. Gutiérrez, *The Power of the Poor in History* (SCM Press, London, 1983), p. 45. For a discussion of the relationship between liberation theology and dependency theory, see W. R. Garrett, 'Liberation Theology and Dependency Theory', in *The Politics of Latin American Liberation Theology: The Challenge to U.S. Public Policy*, R. L. Rubenstein and J. K. Roth (eds) (The Washington Institute Press, Washington, 1988), pp. 174–98.
14. G. Gutiérrez, *The Power of the Poor in History*, p. 78.
15. *Ibid.*, p. 192.
16. L. Boff and C. Boff, *Introducing Liberation Theology* (Burns and Oates, Tunbridge Wells, 1987), p. 30.
17. F. M. Heredia, 'Christianity and Liberation: A Cuban Study of Latin American Liberation Theology', *Social Compass*, vol. 35, issue 2/3 (1988), pp. 309–42, see especially pp. 332–42.
18. P. G. Moll, 'Liberating Liberation Theology: Towards Independence from Dependency Theory', *Journal of Theology for Southern Africa*, vol. 78 (1992), pp. 25–40.
19. *Ibid.*, p. 36.
20. *Ibid.*, pp. 38–9.

21. L. Cormill, 'Correspondence: Lee Cormill Responds to Peter Moll', *Journal of Theology for Southern Africa*, vol. 82 (1993), pp. 88–94.
22. See P. Ekins, *A New World Order: Grassroots Movements for Global Change* (Routledge, London, 1992), pp. 99–100.
23. W. M. Adams, *Green Development: Environment and Sustainability in the Third World* (Routledge, London, 1990), p. 8.
24. For a comparison of red and green strategies of development, see M. Friberg and B. Hettne, 'The Greening of the World: Towards a Non-Deterministic Model of Global Processes', in *Development as Social Transformation: Reflections on the Global Problematique*, H. Addo, S. Amin, G. Aseniero, A. G. Frank, M. Friberg, F. Frobel, J. Henrichs, B. Hettne, O. Kreye and H. Seki (United Nations University, Hodder and Stoughton, Sevenoaks, 1985), pp. 204–70.
25. B. May, *Third World Calamity* (Routledge and Kegan Paul, London, 1981), p. 226.
26. See M. Friberg and B. Hettne, *op. cit.*, p. 220.
27. P. Ekins, *op. cit.*, p. 100.
28. R. Bahro, *Socialism and Survival* (Heretic, London, 1982); R. Bahro, *From Red to Green: Interviews with New Left Review* (Verso, London, 1984).
29. R. Bahro, *op. cit.*, 1984, p. 211.
30. R. Kothari, 'Environment, Technology and Ethics', in J. R. Engel and J. G. Engel (eds), *op. cit.* pp. 27–35.
31. C. Palmer, 'Some Problems with Sustainability', *Studies in Christian Ethics*, vol. 7, issue 1 (1994), pp. 52–62; quotation from p. 53.
32. *Ibid.*, pp. 54–5.
33. *Ibid.*, p. 58.
34. H. Rolston III, 'The Wilderness Idea Affirmed', *The Environmental Professional*, vol. 13 (1991), pp. 370–7.
35. W. M. Adams, *op. cit.*, p. 84.
36. L. Brown, C. Flavin and S. Postel, 'Green Taxes', in *The Earthscan Reader in Sustainable Development*, J. Kirkby, P. O'Keefe and L. Timberlake (eds) (Earthscan, London, 1995), pp. 343–7.
37. H. E. Daly (ed.), *Economics, Ecology, Ethics* (W. H. Freeman Company, San Francisco, 1980).
38. H. E. Daly, 'The Steady-State Economy: Alternatives to Growthmania', in J. Kirkby, P. O'Keefe and L. Timberlake (eds), *op. cit.*, pp. 331–42.
39. A. Naess, 'Sustainable Development and Deep Ecology', in J. R. Engel and J. G. Engel (eds), *op. cit.*, pp. 87–96.
40. S. R. Sterling, 'Towards an Ecological World View', in J. R. Engel and J. G. Engel (eds), *op. cit.*, pp. 77–86.
41. *Ibid.*, p. 82.
42. R. Robertson, 'Liberation Theology, Latin America and Third World Underdevelopment', in R. L. Rubenstein and J. K. Roth (eds) *op. cit.*, pp. 117–34.
43. T. Verhelst, *No Life Without Roots: Culture and Development*, trans. Bob Cumming (Zed Books, London, 1990), p. 158.
44. *Ibid.*, p. 49. See also R. Robertson, 'Liberation Theology, Latin America and Third World Underdevelopment', in R. L. Rubenstein and J. K. Roth (eds), *op. cit.*, pp. 117–34.
45. For a discussion of the integration of the cosmic and the metacosmic in an Asian theology of liberation and its rejection of the Western assumptions of Latin American liberation theology, see A. Pieris, *An Asian Theology of Liberation* (T. and T. Clark, Edinburgh, 1988). Also relevant is S. Ryan, 'The Search for an Asian Spirituality of Liberation', in *Asian Christian Spirituality: Reclaiming Traditions*, V. Fabella, P. K. Lee, D. Kwang-sun Suh (eds) (Orbis, Maryknoll, 1992), pp. 11–30.
46. T. Verhelst, *op. cit.*, p. 50. For additional comment on the relationship between the ethics of culture and ecology, see E. Dussel, *Ethics and Community* (Burns and Oates, London, 1988), pp. 194–204.
47. D. Hallman (ed.), *EcoTheology: Voices from South and North* (Orbis, Maryknoll, 1994).
48. M. Sowunmi, 'Giver of Life – "Sustain Your Creation"' in D. Hallman (ed.), *op. cit.*, p. 153. See section called 'Insights from Indigenous Peoples', pp. 207–27.
49. R. Cooper, 'Through the Soles of My Feet: A Personal View of Creation', in D. Hallman (ed.), *op. cit.*, pp. 211–12.

50. G. Tinker, 'The Full Circle of Liberation: An American Indian Theology of Place', in D. Hallman (ed.), *op. cit.*, pp. 218–24.
51. *Ibid.*, p. 220.
52. D. Curtin, 'Making Peace with the Earth: Indigenous Agriculture and the Green Revolution', *Environmental Ethics*, vol. 17 (Spring 1995), p. 71.
53. L. Boff, 'Social Ecology: Poverty and Misery' in D. Hallman (ed.), *op. cit.*, p. 245.
54. L. Boff and V. Elizondo (eds), *Ecology and Poverty, Concilium 5* (SCM Press, London, 1995).
55. L. Boff, 'Liberation Theology and Ecology: Alternative, Confrontation or Complementary?', in L. Boff and V. Elizondo (eds), *op. cit.*, pp. 67–77. In the same volume Rosino Gibellini makes the point that ecology as a theme in liberation theology has only emerged after the Rio de Janeiro conference in 1992: R. Gibellini, 'The Theological Debate on Ecology', in L. Boff and V. Elizondo (eds), *op. cit.*, pp. 125–34. See also L. Boff, *Ecology and Liberation: A New Paradigm* (Orbis, Maryknoll, 1995).
56. L. Boff, 'Liberation Theology and Ecology: Alternative, Confrontation or Complementary?', in L. Boff and V. Elizondo (eds), *op. cit.*, pp. 74–5.
57. W. E. Hewitt, 'Myths and Realities of Liberation Theology: The Case of Basic Christian Communities in Brazil', in R. L. Rubenstein and J. K. Roth (eds), *op. cit.*, pp. 135–55.
58. See P. E. Sigmund, 'The Development of Liberation Theology: Continuity or Change?', in R. L. Rubenstein and J. K. Roth (eds), *op. cit.*, pp. 21–47. Rubenstein argues that liberation theology leads to the adoption of middle-class Western values. It seems to me that his suggestion that 'under the impact of liberation theology today's poor could become tomorrow's new secularised bourgeoisie' (p. 81) is exaggerated. See R. Rubenstein 'Liberation Theology and the Crisis in Western Theology', in R. L. Rubenstein and J. K. Roth (eds), *op. cit.*, pp. 71–95.
59. R. Kothari, 'Environment, Technology and Ethics', in J. R. Engel and J. G. Engel (eds), *op. cit.*, p. 34.
60. V. Shiva, *Staying Alive: Women, Ecology and Development* (Zed Books, London, 1988), p. 15.
61. R. Radford Ruether, *Gaia and God: An Ecofeminist Theology of Earth Healing* (SCM Press, London, 1992), p. 258. While there is a plethora of books linking feminist theology with ecology, there are fewer which deal specifically with ecology and development issues. See, however, note 63 below. For general books on ecotheology see, for example, A. Primavesi, *From Apocalypse to Genesis: Ecology, Feminism and Christianity* (Burns and Oates, Tunbridge Wells, 1991); S. McFague, *The Body of God: An Ecological Theology* (SCM Press, London, 1993); E. Green and M. Grey (eds), *Ecofeminism and Theology*, Yearbook of the European Society of Women in Theological Research (Kok Pharos, Kampen, 1994).
62. R. Radford Ruether, *op. cit.*, p. 266.
63. R. Radford Ruether (ed.), *Women Healing Earth* (SCM Press, London,1996).
64. E. Green, 'Reflections on Ecofeminist Theology and the Feminist Critique of Science', *Theology in Green*, vol. 5, issue 1 (Spring 1995), pp. 34–7.
65. I. Dankelman and J. Davidson, 'Planning the Family: A Woman's Choice', in J. Kirkby, P. O'Keefe and L. Timberlake (eds), *op. cit.*, pp. 293–7.
66. S. McDonagh, *The Greening of the Church* (Orbis, Maryknoll, 1990), pp. 59–73.
67. J. Moltmann, *The Future of Creation* (SCM Press, London, 1979), p. 5.
68. *Ibid.*, pp. 110–13.
69. L. Boff and V. Elizondo, 'Ecology and Poverty: Cry of the Earth, Cry of the Poor', in L. Boff and V. Elizondo (eds), *op. cit.*, pp. ix–xii.
70. G. Kruijer, *Development through Liberation: Third World Problems and Solutions*, trans. A. Pomeraus (Macmillan Education, Basingstoke, 1987).
71. *Ibid.*, pp. 32–3.
72. M. Taylor, *Good for the Poor: Christian Ethics and World Development* (Mowbray, London, 1990), pp. 47ff.
73. P. Sands, 'Enforcing Environmental Security', in J. Kirkby, P. O'Keefe and L. Timberlake (eds), *op. cit.*, p. 260.
74. *Ibid.*, p. 261.
75. *Ibid.*, pp. 262ff.
76. D. Bull, *A Growing Problem: Pesticides in the Third World* (Oxfam Books, Oxford, 1982).
77. W. M. Adams, *op. cit.*, pp. 117–22.

78. T. W. Hartley, 'Environmental Justice: An Environmental Civil Rights Value Acceptable to all World Views', *Environmental Ethics*, vol. 17 (1995), pp. 277–89.
79. For an analysis of how environmental justice affects environmental policy, see V. Been, 'What's Fairness Got to Do with It? Environmental Justice and the Siting of Locally Undesirable Land Uses', *Cornell Law Review*, vol. 78 (1993), pp. 1001–85.
80. P. Sands, *op. cit.*, p. 263.
81. D. Ghai (ed.), *Development and Environment: Sustaining People and Nature* (Institute of Social Studies, Blackwell, Oxford, 1994).
82. *Ibid.*, p. 8.
83. J. Zizioulas, 'Preserving God's Creation: Part 1', *Theology in Green*, issue 5, (January 1993), pp. 16–17.
84. J. Moltmann, *The Church in the Power of the Spirit* (SCM Press, London, 1977), p. 190.
85. S. McDonagh, 'Did Rio Achieve Anything?', *Theology in Green*, issue 8, (October 1993), pp. 3–12.
86. D. Reid, *op. cit.*, p. 212.
87. J. Roddick, *Scottish Academic Network for Global Environmental Change Bulletin* (University of Glasgow, Autumn/Winter 1993).
88. W. M. Adams, *op. cit.*, p. 148.
89. N. Abel and M. Stacking, 'The Experience of Underdeveloped Countries', in *Project Appraisal and Policy Review*, T. O'Riordan and W. R. D. Sewell (eds) (Wiley, Chichester, 1981), pp. 253–95.
90. W. M. Adams, *op. cit.*, p. 160.
91. C. Watson, 'Working at the World Bank', in *Aid: Rhetoric and Reality*, T. Hayter and C. Watson (eds) (Pluto Press, London, 1986), pp. 268–75.
92. W. M. Adams, *op. cit.*, p. 167.
93. E. F. Schumacher, *Small is Beautiful: Economics as if People Mattered* (Blond and Briggs, London, 1973).
94. W. M. Adams, *op. cit.*, p. 197.

6

Implications for a new science

You might be wondering now: Where do we go from here? How can theology make a positive and practical contribution to environmental concerns in the light of the issues I have raised in the previous chapters? Is there an approach which lends itself to new ways of thinking about humanity in relation to the natural world? I argued in the first chapter that an ambiguity in our response to nature is part and parcel of the Christian story. On the one hand, Christianity points to a mystical contemplation of God which seems to encourage a spiritual ascent, an escape from worldly and material concerns in other-worldly asceticism. On the other hand, the Judaeo-Christian tradition is replete with images of fecundity, an exhortation to adore God in the world around us, a celebration of God's immanence in all and through all. I also argued that this double imagery is not so much a contradiction as a necessary paradox. A distorted image of God and the natural world emerges if either is emphasized at the expense of the other. While the former on its own leads to alienation from nature, the latter leads to identification with nature.

But what about the ambiguous attitude to the natural world implicit in the practice of science and technology? In this case I suggested that such ambiguity has negative consequences. On the one hand, a culture of science seems to create a security within which humans can live with impunity. However, this is a partial delusion, since those very structures are subject to flux and change in a way which is largely hidden from view. On the other hand, a culture of technology encourages frenzied and intense activity, an ever grasping thrust towards the new, an arrogance born of the failure to 'let things be'.

The overall confusion resulting from this twin message of security at the price of frenzied activity has encouraged public anxiety. Robin Grove White suggests that the lack of trust in the authority of science is part of a wider cultural rejection of any authority structures.[1] This includes the authority of the state, the authority of politics and the authority of the church. As a consequence, environmental policy which is couched in the language of authority simply will not work. Even the non-government organizations (NGOs) which seem to work from the 'bottom up' are becoming too institutionalized and identified with government officialdom for public resonance. Any environmental policy, and it seems to me any theology, which is in tune with the current cultural mood has to take into account the 'growing proliferation of frequently fragile, but vibrant social networks developing around practices concerned with such issues as health, food, gender, personal growth and spiritual development, counselling, leisure, animals, co-ops of all kinds and new religious networks in the cities'.[2]

What language shall we speak?

I suggested in the first chapter that science could not have emerged unless the natural world was thought of as an object. To put it another way, one of the preconditions of the emergence of science is the appearance of a language which is disjunctive, rather than conjunctive.[3] By disjunctive I mean it allowed for a difference between the language and the reality to which that language referred. This growing sense of the representational nature of language gradually emerged between the twelfth and sixteenth centuries. This contrasts with the classical and medieval era where language is conjunctive, there is no distinction between the world of thoughts and words and the things. In this scenario analogy and metaphor dominated, rather than analysis and induction, characteristic of the disjunctive view of language. Once language is part of a human construct, there is no guarantee that language and reality will be coincident. Following from this disjunctive language the relationship between humanity and nature shifts. This is not simply an affirmation of difference, which is characteristic of anthropocentric attitudes stemming from the early cultures of the Graeco-Roman period, including Judaeo-Christian traditions, but a stress on 'I–It', not 'I–Thou' relation.

Such language fosters an attitude of domination and control. Whenever the Other becomes an It, this encourages exploitation and manipulation of the Other for self-centred ends. The experience of the oppressed, whether through race, culture, age, gender, disability, class, cast, to name a few examples, is the experience of being treated as an It. A commodity that can be disposed of once 'its' usefulness is over. Even in so-called advanced societies, such as our own, the experience of being an It can be felt by those suffering under insecure employee–employer relationships characteristic of the 1990s. What used to be thought of as a source of security no longer delivers the goods. I am arguing here that it is part of common experience to have some knowledge of what it is like to be an It. Now whether animals, plants and the like have any such sensitivity to this experience is a moot point. I suspect St Francis's empathy with the natural world, which arose out of his affirmation of the whole world as Thou, is a rare commodity today. Indeed his world was replete with metaphors expressed through conjunctive language. The natural world for him was his world, a text from which to draw conclusions about the wonder of the Creator.

If the modern problem is this gulf between language and reality, then science now attempts to bridge this gulf by assuming that the language it uses is an adequate and transparent representation of the things it describes. Yet it still assumes that the thing in itself is extra-discursive reality, word and world no longer part of one symbolic order. This is taken further in moral discourse, where 'in the late seventeenth and eighteenth centuries a gulf between language and moral realities comes to seem self evident, as ethical discourse takes on the disjunctive characteristics of modern science'.[4] The natural world is the world of 'facts', according to such a view it is incapable of giving humanity any clues about our place in the world. Humanity seems to become an autonomous observer of reality. Szerszynski argues that all such attempts to ground ethics in an extra-discursive reality constructed by human beings is bound to fail. He suggests:

> Firstly, the notion of a reality beyond language, to which language refers, is a product of our own particular modern understanding of language, and is thus not available as something in which language can itself be grounded. Secondly, the ideal of a formalised language completely abstracted from the social and cultural is itself impossible to achieve.[5]

Ecological ethics seems to be able to get round this problem by recognizing ourselves as part of the world; the ethical laws now emerge from the natural world. But this view still retains the modernist assumption that there is a close correspondence between truth and reality. It seems that environmentalism which draws on this idea has not taken into account neo-modernism which acknowledges that all forms of knowledge are imperfect, while still assuming the ideal of progress. The emphasis for neo-modernism is not so much content, but process. The practical result is to set up a dynamic which is assumed to lead to a desirable goal, rather than deliberately aimed at for a desirable end. Democratic approaches to the environment assume that it is through use of language itself in discussion and debate that intrinsic rationality is generated.[6] Habermas argues that the environmental movement is valuable, not so much for the content of its arguments, but because of the communicative spaces it opens up.[7] But the logical outcome of the shift from content to form is a grim one: 'a lonely humanity faced with the task of pure self-assertion in a meaningless world which no longer tells them/us what to do'.[8]

Has deep ecology discovered a solution to this problem by refusing to accept a subject-object dualism? I argued earlier that a re-embeddedness of humanity in the world in a way which tries to ignore any distinctions leads to a different kind of alienation. On the one hand, the cosmic whole seems to deny individual value, while on the other the search for meaning leaves self isolated and alienated from others. Szerszynski has noted the same trend in expressivism and is not far off the mark when he states that:

> Existence becomes 'hollowed out', with meaning and purpose residing only at the extreme ends of reality – in the self and in the cosmos only. Between, there is just meaningless accident. Stoicism – and deep ecology – turns out to be just another, disguised version of gnostic alienation.[9]

The question now becomes: How can we recover a conjunctive language without denying our specific cultural context? A search to capture a 'lost paradise' seems fraught with difficulties in that it seems to ignore the present realities which we face day to day. The language that we need seems to be one which fosters what Heidegger describes as our experience of being in the world as 'ready-to-hand', through our ongoing action.[10] This contrasts with the scientific view of the world which sees human–world relationship as 'present-at-hand', a set of objects controlled by the detached observer. I have suggested earlier that the practice of science is also one which is in part 'ready-to-hand', in that it is also a process of discovery through the work of the whole community. Nonetheless, the public image of science is one that is 'present-at-hand' and it is this public image which leaves the most lasting impression.

Exploring the wisdom motif in Russian Orthodoxy

The following three sections argue the case for using wisdom as a means through which the conjunctive language that we need can begin to be fostered and nourished. Furthermore, such a term can become inclusive of science, as I suggest below. The practical outcome of such a view will become apparent in the sections which follow.

One example of wisdom in 'folk' religious experience is that rooted in popular ancient Orthodoxy, as expressed through mosaics and icons dedicated to wisdom.[11] Yet the search for wisdom goes much further back than this: it is a motif as old as the tradition of philosophy. Aristotle sought after Sophia as that which bridged scientific knowledge with intuitive reason in seeking those things which are 'highest by nature'.[12] He considered that the principles Sophia grasps intuitively (*nous*) are theological, but the things flowing from these principles are known through the element of scientific knowledge (*episteme*). The two independent sciences emerge as theology and ontology, with Sophia identified in part with both areas. Aristotle's search for Sophia as that which bridges the gap between God and creation became an element in the theological traditions of the early church fathers and developed into a full-blown Sophiology of later nineteenth-century Russian Orthodox thought. Both traditions sought Sophia as a way of interpreting biblical wisdom found in the Old and New Testament. The Hebrew concept of wisdom, which began with the practical instruction of the sages, later evolved into a concept that allowed it to be identified with the Spirit of God.[13] In this way the universal cosmic principles characteristic of Greek philosophy and the early common sense wisdom of Hebrew sages could be brought together.

I will continue this section with a brief reflection on the work of Bulgakov as an example of Sophiology in Russian Orthodoxy, especially in the context of themes relevant for ecotheology. I will then move on to consider how Sophia is becoming part of feminist theology, before offering my own particular interpretation in the light of my own particular interest in developing an ecotheology.

Sergius Bulgakov (1871–1944) had mystical experiences which fed into his theological writing. He believed that God created the world on the foundation of wisdom, which was common to the whole Trinity. The 'ideal forms' of creation are present eternally in God, rather like divine prototypes. His book *The Unfading Light*, published in 1917, developed the idea of the cosmic Sophia as the intelligible basis of the world, the wisdom of 'nature'.[14] Creation is viewed as a theophany: a manifestation of God through Sophia. Bulgakov argued that the legacy of the Reformation and the Counter-Reformation is a refusal to consider the link between cosmology and anthropology. I believe that this is an important insight in that there is a tendency to stress anthropology as a reaction against medieval cosmologies: history replaces creation as the paradigm for theology. He insisted that Sophiology is a dogmatic interpretation of the world that has a rightful place alongside other traditions such as Thomism and Barthianism. In other words, Sophiology is a way of interpreting all theology, not just a theology of creation. For Bulgakov:

> The future of living Christianity rests with the Sophianic interpretation of the world and of its destiny. All the dogmatic and practical problems of modern Christian dogmatics and ascetics seem to form a kind of knot, the unraveling of which inevitably leads to Sophiology. For this reason in the true sense of the word, Sophiology is a theology of *crisis*, not of distinction, but of salvation.[15]

It is this salvific role ascribed to Sophia which unnerved more traditional theologians of this era since it seemed to replace the role of Christ.

His doctrine of the Trinity was also a subject of controversy. He believed that the being (*ousia*) of God should not be separated from wisdom-glory, where wisdom signifies content and glory manifestation of all persons of the Trinity. He rejects, then, any notion that wisdom is confined to the *logos*, since this would suggest the Father and Spirit are without wisdom. Wisdom is the characteristic of all three hypostases, just as the divine *ousia* is shared by them all. While his claim that Sophiology is the key to the future of Christianity strikes a modern ear as somewhat exaggerated, once we accept that Sophia is part of the being or *ousia* of God, then it seems to me that it is inevitable that Sophiology and salvation are intimately connected. Sophia is the means through which the divine ideas become reality, so that the life of truth, identified with the *logos*, when it becomes fully transparent is *beauty*, identified with the Spirit, and hence reflecting the divine glory. To cite Bulgakov here: 'The life of Truth in its full transparency is beauty, which is the self-revelation of the Deity, the garment of God, as it were, it is the divine glory that the heavens declare.'[16] It is this beauty which seems to be the mark of salvation, and as such shows the involvement of all three persons of the Trinity.

For Bulgakov Sophia is the 'eternal feminine' of the Godhead, becoming both the internal love of God and the connection between the created world and the eternal world of the triune God. Divine Sophia becomes creaturely Sophia in the creation, expressed because of the love of God. Sophia in her creaturely and divine modes is expressed through the joint action of Word and Spirit, again pointing to the love of the Father rather than themselves. In this way he has still clung to the primacy of the action of the Father in creation, but it is through participation of Word and Spirit. The eternal Son becomes the *logos* of the world and the Spirit becomes the beauty and glory of the world created though Sophia as an expression of God's love. The Spirit is present as creaturely Sophia at the beginning of creation before the first 'Let there be' of God's creative act. She is like the first mother who brings forth life to all that exists in the created world. I find the link which he develops between Sophia, the *logos* and the Spirit fascinating. The bringing forth of life in creation becomes a trinitarian act, not just 'Mother' in place of 'Father', but a movement of trinitarian love through Sophia. This coheres well with my own theological stance, to which I will return later.

Bulgakov was anxious to avoid charges of heresy; nonetheless his ideas were subsequently severely criticized.[17] While aspects of Bulgakov's theology may need to be re-examined, it is primarily his refusal to see the *ousia* of God in abstract terms that marks out both the novelty and controversy in Bulgakov's theology.[18] The contemporary Orthodox theologian John Zizioulas has argued that God's Being and the Being of the Church are communion.[19] Why not describe this communion of love as Sophia? Surely love and wisdom are counterparts in the *ousia* of God?

A more serious criticism of all Sophiological schemes is whether there is sufficient consideration given to evil present in the universe. However, Bulgakov, at least, does devote some space to this issue. The counterpart to the creature is 'creaturely nothingness' which is the 'dark face of Sophia'.[20] Bulgakov is able to achieve this in a way that was impossible for Solovyov, since he separates creaturely Sophia from divine Sophia. The 'shadow' side of Sophia is in her creaturely aspect.

Wisdom in feminist theology

The feminist approach to the theme of wisdom, in contrast to the Sophiology discussed above, takes as its starting point the experience of wisdom in the world, rather than wisdom as a theological term which links with creaturely wisdom. While there are links between certain aspects of Bulgakov's Sophiology and feminist thinking, as I have tried to show above, the experience of God in Sophiology is through mysticism rather than through concrete modes of thinking. The overall result, namely a reflection on Sophia as feminine in the world comes from 'above', rather than 'below'; yet it seems to me that the language is the same: namely the language of love and entreaty, even if the direction is rather different.

Catherine Keller has drawn on the idea of emancipatory wisdom as that which best describes the future of theology in the university.[21] It is wisdom that can straddle the world of the academic and ecclesiastical communities to which theology must give an account of itself. For Keller, wisdom 'at least as practised in the indigenous and in the biblical traditions, is irredeemably implicated in the sensuous, the communal, the experiential, the metanoic, the unpredictable, the imaginal, the practical'.[22] It differs from the coercive control of matter by mind, which is the subject of modernity; rather it takes time to 'let things become' and include the social in the cosmological. This echoes the thought of Heidegger which I discussed in the introduction. Keller insists that the biblical notion of wisdom was linked to principles of social justice for the poor and vulnerable in the community. These principles prevent it from becoming a version of spiritual individualism, instead it is a source of 'emancipatory imagination'. She believes that wisdom as a source is a movement away from orthodoxies to orthopractice, building on an experiential ground for its claims of truth. It seems to me that Keller's link between wisdom and justice is highly significant as a basis for creating a theology of creation, as I will show later.

Elizabeth Johnson has undertaken by far the most thorough theological treatment of the theme of Sophia from a feminist perspective in her book *She Who Is: The Mystery of God in Feminist Theological Discourse*.[23] She redescribes all aspects of the Godhead through Sophia: namely the Spirit Sophia, Jesus Sophia and Mother Sophia. She argues against confining the feminine in God to the person of the Holy Spirit, which has been the approach of traditional orthodoxy. Her criticisms stem from the fact that the Holy Spirit remains the somewhat elusive 'faceless' person of the Trinity so that we end up with two masculine images and an 'amorphous third person', who as third person is inevitably subordinate.[24] A further difficulty is that stereotypical female roles are projected onto the person the Holy Spirit, thus matching maleness with transcendence and implied superiority and femaleness with immanence and inferiority. She insists that speech of God in female metaphors does not imply feminine dimensions in God as revealed by women, rather:

> Female imagery by itself points to God as such and has the capacity to represent God *not only as nurturing, although certainly that, but as powerful, initiating, creating-redeeming-saving and victorious over the powers of this world*. If women are created in the image of God, then God can be spoken of in female metaphors in as full and as limited a way as God is imaged in male ones, without the talk of feminine dimensions reducing the impact of this imagery.[25]

Her criticism of a readiness to identify the Holy Spirit alone with femaleness and by implication 'nature' and subsequent subordination seems to be well-placed. Her strong reimaging of the whole concept of femaleness, as one which seeks to liberate from stereotypical images of the past where femaleness is identified only with maternal capacities is important for a theology of creation, as I will discuss later.

I particularly welcome her sense of the cosmic role of the Holy Spirit, which counters a more traditional Catholic and Protestant concentration on the Holy Spirit and the church. A sense of the cosmic function of the Holy Spirit is, nonetheless, also characteristic of Eastern Orthodoxy. The cosmic function is important in a theology of creation as it shows a way of expressing God's immanence in the world without returning to pantheism. While Johnson has envisaged a much broader image of the Spirit, she seems to be using Sophia as a vehicle for expressing the femaleness of the second person of the Trinity. For Jesus Sophia she looks to images of Christ as the wisdom of God in biblical passages such as 1 Corinthians 1:24. She sees this tradition as giving Christology a cosmic dimension, so that wisdom today can function rather like the *logos* metaphor in early Christian centuries. She believes that wisdom metaphors in biblical theology speak of the 'power of relation' which comes to fruition in Jesus as Sophia incarnate.[26]

Johnson's image of Mother Sophia makes explicit the connection with the wisdom tradition: 'Holy Wisdom is the mother of the universe, the unoriginate, living source of all that exists. This unimaginable livingness generates the life of all creatures, being itself, in the beginning and continuously, the power of being within being.'[27] She is like a generative source of all that is, including, it seems, Spirit in the world. The maternal image points not just to God's immanence, but to a new understanding of God's transcendence. A mother's existence in profound, lasting relationships, her life unknown to the child, all point to the hiddenness and transcendence of God.

It is Johnson's insistence on the total replacement of all images of Father by Mother and Son by Child which leads to the removal of any idea of maleness in God. As I will discuss in more detail below, this seems to imply matriarchy in place of patriarchy. While I am in favour of using Sophia as a means of reimaging the Trinity, I would prefer a transforming role of Sophia, so that she becomes the *feminine face* of God. This need not amount to a 'dualistic' stereotypical notion of male and female attributes, which Johnson rightly rejects. Rather, I accept her idea of a transformed feminine, but it is one which then transforms the Fatherhood of God and the Sonship of Jesus as well. This unity in distinction is the dynamism of love in the Trinity, as expressed through Sophia.

Claudia Camp writes on wisdom from rather a different perspective drawing primarily on the Old Testament book of Proverbs.[28] Her purpose is to examine the links between the figure of wisdom as female and wisdom in Proverbs. She believes that those authors who have identified a female goddess with the figure of wisdom have failed to take account of the feminine roles characteristic of the culture of the authors. Her book is wide ranging, drawing on literary, historical, canonical and sociological contexts of wisdom. Her particular question relates to the significance of the femaleness of wisdom, in other words what kind of ideas are brought to bear by the focus on woman.

She points to both the human wisdom tradition as well as the idealized embodiment of all wisdom, demanding commitment to the One who is the source of all Wisdom. The wisdom tradition itself, rather than the sages and kings, became the locus of revelation. Wisdom becomes the mediator between bios and cosmos, the particular and the universal, so that 'the function of the king with respect to nature is not simply replaced but transformed by the language of love'.[29] Her work is of relevance to the present discussion as it shows the rootedness of the biblical wisdom tradition in the experience of women. Moreover, the idea of wisdom as a mediator between bios and cosmos is part of biblical theology and gives some theological justification for Sophiology.

Katherine Dell is one of the few theologians who have attempted to explore in depth the significance of the wisdom tradition in relation to contemporary creation theology. However, she has largely confined her discussion to the notion of 'deep ecology' and how this coheres with the Old Testament themes of wisdom, especially that in the books of Proverbs, Job and Ecclesiastes.[30] She suggests that deep ecology involves a shift from science to wisdom, that is wisdom of the earth or *ecosophy.* She argues against the philosophy of deep ecology which implies that there is no ethical distinction between humanity and other creatures. While this criticism is valid, it seems to me that this is *the* distinctive feature of deep ecology. As I discussed above, 'deep ecology' actually leads to a sense of homelessness, caught in the space between individual expressivism and cosmic unity. Other characteristics which she mentions and affirms in deep ecology, such as the sustaining of life and the interrelatedness of all natural processes, are characteristics of ecology as such, not just deep ecology.

I would also prefer to use as a starting point the more general use of wisdom which is inclusive of science, rather than excluding science. Thomas Aquinas, following Aristotle, was one of the classical theologians who allowed wisdom to become embedded in his theology. This wisdom portrays theology as a science, but more than science as it includes a wisdom which reaches beyond its precepts to critical reflection. In other words 'while *intellectus* is the habit of first principles, and *scientia* is the habit of conclusions, *sapientia* is concerned with both principles and conclusions, and is somehow both *intellectus* and *scientia*'.[31]

Finally, I would like to comment on how Sophia seems to be a relatively new theme for ecofeminism. As I have noted in earlier chapters, Rosemary Radford Ruether is a prominent theologian who has drawn on Gaia as a metaphor for her ecotheology.[32] The advantage of this image is that Gaia has roots in the goddess of the earth, generating an obvious link with the natural world. However, while this is a strength, from a theological perspective it also has its drawbacks. One is that Gaia can seem to become confined to this world, especially from the metaphor generated through James Lovelock's use of this imagery.[33] While some feminist theologians would welcome the return of the goddess and images which stress pantheism, it seems to me that a metaphor which also gives space to express the transcendence of God retains an important part of Christian tradition. This need not culminate in monarchical images or patriarchy, but show the otherness of God as well as the intimacy of God in creation. Sophia is a metaphor which allows such a paradox to become expressed, allowing for distinctions without 'dualism', or more importantly perhaps, a conjunctive rather than disjunctive language.[34]

Wisdom as a metaphor for an ecotheology

The purpose of this section is to show some of the ways in which the different aspects of Sophia are of particular relevance to ecotheology. In general I hope to show how Sophia is a means through which we can arrive at a deeper understanding of the relationship between God and creation by drawing together insights from Sophiology and feminist theology.

Wisdom provides the link between the secular and the sacred. The human search for wisdom, characteristic of secular human longing, becomes transformed into a personification of divine wisdom, as in Proverbs 8. Note that I have not included a detailed discussion on the theme of wisdom in the Old and New Testaments – this would be the subject of another book! Wisdom insists that the world offers insights into the nature of reality, yet looks beyond this to the divine ideal which criticizes all wisdom that is against the commandments of God. As such the juridical role of wisdom is an aspect which can easily slip away from theologies of creation which concentrate simply on the natural realm as their basis. Furthermore, it seems to me that the sense of the transcendence of God is preserved in a more obvious way compared with other theologies which link God and creation, such as through understanding the world as God's body.

Wisdom provides a powerful motif which can stress the importance of the role of the Holy Spirit in creation. The Sophiology of the Russian church presents a strong insight into the way wisdom can serve to shape not just the world of creation, but the inner trinitarian world of the three persons in community. While I would resist a wholehearted adoption of Bulgakov's Sophiology, I believe that Sophia is highly significant for ecotheology. Sophia is the means through which the transcendence and immanence of God is preserved. She becomes the link between God and the world, while retaining the distinction between God and creation. Sophia is not a hypostasis, but the essence of all three hypostases.

Wisdom offers a reminder of the means by which the poetic and rational can come together. The tendency to systematize in purely rational schemes can lead to a loss in the poetic quality of wisdom, which is also fundamental to creation itself.

There is a distinction between the 'natural wisdom' of ecology and the human 'cultural wisdom' of the sages of the Old Testament. However, while there is a distinction between the two ideas, the split between nature and history is not as sharp as it may seem at first sight.[35] Wisdom is characterized by a unitary world in which the regularities in the human and historical-social realm are no different, in principle, from those in the non-human realm. This implies that an ecotheology needs to become inclusive of the historical sphere, rather than existing simply as a backdrop to the human drama of history.

Wisdom offers a realistic approach to the theme of creation. Bulgakov recognized that Sophia has a 'shadow side' which faces up to the evil in the universe. In addition, the cross becomes the wisdom of God in the New Testament. It is through this wisdom that creation and redemption are brought together in a single theme, rather than separated as tends to be the case for much contemporary creation-centred spirituality. It also highlights the suffering of God with creation in a paradoxical way. Once the wisdom of all of creation is seen as caught up in the wisdom of God as expressed in the cross, the cross becomes the place of crucifixion of all of creation, not just suffering humanity. The image of the *Christa*, so shocking to many, is complemented by the radical image of crucified nature.[36] Here nature is an inclusive term: humanity in solidarity with all creation. While leaves and flowers have traditionally been associated with the theme of the resurrection, in our ecologically stunted age, dying plants can come to express crucifixion as well.

Wisdom provides a means of reaffirming the feminine in the Trinity.

I would tend to agree with Elizabeth Johnson that the identification of the feminine exclusively with the Holy Spirit stereotypes the femininity in God. The difference between the feminist interpretation of Sophia and Bulgakov's Sophiology seems to be that for Bulgakov Sophia becomes the feminine essence of the divine Trinity understood in more traditional ways. However, Johnson uses Sophia as a metaphor to redescribe the divine Trinity, not by including a feminine element, but by a redescription of God as She who shakes off all previous stereotypes of femininity and patriarchy. The problem now is how can this powerful, female image coexist with the traditional male image?

Johnson is careful to suggest that her theology is a deliberate corrective to other metaphors which have seemed to presuppose male imagery. The difficulty with this approach is that it would seem to portray God in a way that is incompatible with traditional views. If it is a corrective, surely this implies that traditional views have their place as well? While God as feminine needs to be inclusive of all the strong characteristics that Johnson suggests, I prefer the idea that Sophia can become the feminine divine transforming all three persons of Father, Son and Spirit. In other words Sophia is the feminine face of God. In speaking of the face of God I am quite deliberately not going down the route of Goddess imagery.

As I mentioned earlier, Rosemary Radford Ruether has used Gaia as her way of including the feminine divine in traditional images of God. However, because of the strong influence of Gaia as goddess in much contemporary thought, not to mention the Gaia of James Lovelock, I prefer to use Sophia. Sophia, unlike Gaia, has some basis in Old Testament imagery, especially the Proverbs and the Wisdom of Solomon. Here she is portrayed as the divine worker in creation.

Wisdom extends the discussion of the role of all three persons in creation. In speaking of Sophia Father I am implying that God is also Mother who moves beyond the realm of sexual differentiation into divine mystery. Sophia in all three persons works together in a unity of communion and love, so that it is wisdom as divine glory which radiates from the immanent Trinity. With regard to an ecotheology, Sophia highlights the role of the feminine divine in the creative process. Sophia as Mother encourages the concept of wisdom as one who gives birth to the world. However, as I indicated above, I would also wish to retain the idea that wisdom transforms the role of God as Father. This bears some resemblance to Jürgen Moltmann's idea of the 'Motherly Father'.[37] However, I would wish to argue for a Mother and Father in God, that is God showing the characteristics of both metaphors. While moving closer to E. Johnson's position, by combination of both metaphors in a mystery of divine being it avoids the two possible difficulties of either replacing the idea of God with exclusive female images, or being content with a feminine aspect in God which does not do justice to the strength of the female metaphor. The fatherhood of God only makes sense in the light of the feminine divine as expressed by Sophia. In this way God as Father of creation is not seen as a threat, but as an overflowing of divine love.

While wisdom from the Russian Orthodox perspective tends to stress the divinity of wisdom, beginning with a reflection of wisdom as mystical experience of the divine, feminist theological approaches stress the creaturely aspects of wisdom. It seems to me that while these two approaches are in some respects similar, in that they both see wisdom as a means of linking God and creation, they serve to complement each other. The danger of the Orthodox approach is that wisdom can become a static ideal form. The possible tendency of feminist approaches is that wisdom remains earthbound, subject to contingency where God's immanence is stressed at the expense of transcendence. It seems to me that by keeping both ways of thinking in view a genuinely holistic approach to a theology of creation can be explored.

The future of science through the lens of wisdom

What kinds of futures are predicted by scientists and how is this challenged by the notion of wisdom? Watts discusses the implicit eschatology of science and offers an interpretation of science as salvation.[38] I agree with Watts that hope in science is an important motivation and driving force for many scientists. Yet I am concerned that he seems to suggest that scientific hope and theological hope are counterparts. The distinction between the two comes out clearly in Jürgen Moltmann's discussion of heaven.[39] Moltmann is critical of Marx's interpretation of heaven as idealized society. Such a heaven is one without alternative possibilities and collapses in the wake of communism, that is the fulfilment of these wishes. His criticism of secular ideas of heaven resemble his criticism of secular ideas of hope, which he covered in his earlier works.[40] As before, he is more favourably disposed towards Bloch's work. Bloch considers that heaven is not just a historical future, but an eschatological future.[41] Bloch distinguishes between subjective hope, *qua sperator*, and the object of hope, *quae sperator*. According to Bloch, we can only have confidence in the former 'hoping hope', not the 'hoped hope'. If we had a guarantee of the latter objective hope, our hoping would be 'trivial', rather than 'brave'. This leads on to the idea of an 'empty space' in which hope stands, which Moltmann brings into his discussion.[42] Bloch's refusal to make objective hope certain weakens the ontological content of his idea of 'Not Yet' and links it in a closer way with history. However, he also pressed towards a concept of the 'dialectical leap into the new' and resisted any thought that this was 'preordered' 'in the style of the old teleology', let alone a teleology 'mythologically guided from above'.[43] Moltmann distances himself from Bloch's atheistic description of heaven, believing that a 'transcending without transcendence' ultimately slips into an 'indefinite endlessness'.

Moltmann seems to be anxious to preserve a Christian interpretation of heaven, over against secular alternatives which appear to remove the necessity for God. His lengthy discussion on heaven in Bloch's philosophy reflects his own desire to transform this philosophy into Christian categories. He is especially critical of Bloch's seeming identification of heaven with the kingdom of God. The overall conclusion from this discussion is that all promise from science is inevitably a promise from '*within*'. In other words, salvation from within the boundary of science is not equivalent to theological salvation. For Moltmann, to equate the two actually leads to atheism since there is no longer any need for a transcendent God. In addition, release from ignorance, which Watts parallels with religious release from sin, in theological terms seems to be a strange analogy. The archetypal account in Genesis seems to equate sin with loss of ignorance, leading to a progressive deterioration in human relationships with God, with each other and with creation. Fallenness seems to be equivalent to the progressive thirst after knowledge, power and independence from God.

Watts also tries to find parallels between artificial intelligence and eschatology.[44] Even if the goal of replacing the human mind is achieved, we cannot expect the computer to go beyond the human intelligence. While it could be argued that the goal of artificial intelligence reflects a vision of the future, this is true of all science, from biotechnology to nuclear physics. Is the potential impact of this greater than other sciences? The perception of humanity as just a function of intelligence, which is the view of 'strong AI', is something to be resisted. It seems to me to raise all kinds of ethical questions about value and the relative value of animals and material creation. How far are such beings, even if they could be created, really life forms? If not, do they have moral standing or significance? Furthermore, such an approach ignores the human quality of wisdom, which goes beyond the artificial intelligence defined in terms of the rational capabilities of humankind. Frank Tipler's book, *The Physics of Immortality*, has come under some heavy scrutiny from other scientists. Chris Clark, who is professor of mathematics at Southampton University, makes some pertinent remarks:

> By 'life' Tipler means the AI concept of information processing. He argues, quite plausibly, that for this to continue for ever in the future it is necessary for life to colonise the entire universe as a global super-intelligence able to control the dynamics of the cosmos in detail. He assumes, for no obvious reason, that it is *Homo sapiens* that will initiate the process, by sending into space probes loaded with the genetic information necessary for reproducing humanity throughout the universe – *continuing into space the domination of the physical world that has currently brought us to the brink of destruction on earth. This science fiction scenario underlies the real motives behind this book, about which Tipler is quite explicit. They lie not in the pursuit of a greater truth that embraces both science and religion, but in proclaiming his faith that ultimate truth, even God, is to be found in the Physics that has been formulated over the last ten years, a physics rooted in the fields and particles that ignores all human experience of spiritual reality.*[45]

Richard Dawkins's biological thesis is perhaps closer to the practice of environmental biology. However, his theories make explicit the inherent agenda of much mechanistic philosophy, namely that the future is one which is ultimately selfish, rather than altruistic.[46] Richard Lewontin, on the other hand, is sharply critical of any attempt to link genetic blueprint with social effect. He is a population geneticist who is concerned with the tendency in biology towards forms of absolutism. He argues that biologists who suffer from 'physics envy' miss the point as there are no absolute laws in biology. He believes that the reason behind seeking such elegant explanatory causes is: 'They are looking for simple causes and general phenomena, because that's how you get rich and famous. You don't get rich by saying "everything depends".'[47] The biological determinism that is part and parcel of the philosophy of biotechnology draws on this simplistic view of biology which Lewontin rejects. It seems to me that in saying 'everything depends', Lewontin is integrating the theme of wisdom into his philosophy. Dawkins's ideas may be logically elegant and mathematically ingenious, but they lack wisdom. This is illustrated further by his reduction of religion to a scientific theory in competition with other scientific theories, which seems to me to be 'reductionism' *par excellence*, but without any attempt at justification.[48]

In contrast to these debates within biology, Lovelock's Gaia theory seems attractive to those who are seeking a philosophical basis for a more altruistic world view, by stressing cooperation, rather than competition. As I mentioned earlier, Sahouris argues that Gaia can become the basis for moral value in the world, by seeking for a holistic, interdependent and corporate world.[49] The future of the earth now is the survival of life, where life is defined as biota, the sum of life on earth. Lovelock strongly denies any teleology in his theory, but some of his followers are less cautious and see the future as one in which ultimately Gaia survives, even at the expense of humans.

This combination of science and mythology, drawing on the theme of the ancient goddess of the earth, conjures up a powerful image that is a striking counterfoil to the dream of artificial intelligence with its presumed anthropomorphic arrogance. Nonetheless, I am deeply suspicious of the future scenarios of either pseudo-scientific myths. The supposed life of artificial intelligence programmes is as alien to me as the survival of micro-organisms predicted by Gaia since these are the only organisms which ultimately contribute to keeping the environment constant.[50]

Does science offer an alternative eschatology? While it is true that in the wake of secularization the void opened up encouraged visions of the future emerging from science, the shape of these visions and the images they portray bear very little resemblance to Christian eschatology. In some cases, as in the more political dreams, the image collapses once the dream is realized. In other cases, the ultimate logic of the projected future is a bleak, sparse landscape, which has little place for human life, religious life or any other life as we know it. It is only when some elements of these ideas can become incorporated into Christian eschatology, stemming as it does from the significance of the Christological event and looking to the future in God, that science will be able to make a positive contribution.

Wisdom as a paradigm for environmental ethics

How might a metaphor of wisdom transform the language of environmental ethics and ultimately the practice of science? Some contemporary philosophers have urged us to move away from the traditional anthropocentric view towards one which gives animals greater value, advocating the endorsement of 'animal rights'. Others have tried to persuade us that it is not animals alone, but all life that is sentient that should be given value. This move to a biocentric approach is paralleled by a move from individualism to community to holism, culminating in giving value to the whole earth system as portrayed in the Gaia myth. One of the problems of a biocentric view is that when combined with holism it leads to locating value in the earth as a single system in a way which seems to lead to the loss of individual value, the loss of the personal. Frederick Ferré has argued that we need a personalistic organicism, which is at the line of tension between 'heedless anthropocentric individualism' and 'mindless biocentric collectivism'.[51] He suggests that those who argue for organicism do so on the basis that personalism displays human arrogance and pride. Furthermore, due respect should be given to the combination of intrinsic and instrumental value in all life forms: instrumental value counts as well. He suggests that personalistic organicism 'may well offer a paradigm for the avoidance of dualisms and dichotomies that have too long plagued environmental philosophy and philosophy in general'.[52] Such an approach allows for distinction rather than separation of humanity and the natural environment.

I argued earlier that distinction is, in fact, necessary for adequate relationship, rather than egalitarianism, which refuses to recognize such distinctions and ultimately leads to paralysis in terms of practical ethics: we no longer know what to do. Ferré's approach resists either/or thinking in favour of both/and thinking. His approach is neither materialist nor idealist, affirming both instrumental as well as intrinsic value. It seems to me that wisdom is another metaphor which comes close to the kind of personalistic organicism that Ferré suggests. Wisdom is a profoundly human quality, yet it is also a cosmological term, reaching out to see wisdom in all things, even the 'lowliest' of species. I believe that Robin Grove-White is right to insist that we need a richer understanding of human identity in order to begin to tackle some of the complex issues of the environmental crisis.[53] The advantage of wisdom as a heuristic tool is that it isolates that human quality which is needed in the sifting process of discernment.

Furthermore, wisdom recognizes the value of experience, that 'right judgement' comes with practice and patience. In this sense it links in with the pragmatic environmental philosophy that is currently in vogue. It also affirms value in the human community which gives due respect to women and men, those marginalized through differences of culture, race or religion. Furthermore, it is suggestive of an ecumenical approach in different religions: the global ethic which Hans Küng advocates needs to be qualified by recognition of the identity of each culture and religion.[54] One of the problems of Küng's approach is that it seems to suggest that a global ethic emerges from a common denominator approach. However, a paradigm of wisdom allows for wisdom within a culture as well as transcultural affinities. Such affinities need not have the last word, but they help in appreciating our common ground. The unity in wisdom is a poetic one, rather than a monistic one which reduces all differences. It seems to me that this attitude is vital in approaching insights from other faith traditions. It helps us to cope with the paradox while we are searching for unity; it is one where distinctions are respected and cultivated. Wisdom refuses to be arrogant, but affirms the highest quality in human nature, wherever this is found. Furthermore, resources of wisdom are as likely to come from Eastern and Southern traditions as from Western ones.

Instead of reason as the all-defining category for humans made in the image of God, which Aquinas suggested, now wisdom is elevated, but this is still qualified by the notion of the Wisdom of God which is known in the fullest extent in Christ's suffering and resurrection. Wisdom as a theological term qualifies all attempts at finding a purely human basis for the construction of environmental ethics. It differs from personalistic organicism in an important respect; namely that it carries an element of surprise, of discovery, the 'eureka' of reaching out and searching for hidden treasures that is also part of the quest for knowledge in science. It takes into account the future, but it is a future 'from ahead', that is an eschatological term. It seems to me that a redefinition of the future in science is fundamental a priori in our perception of ethical action. In other words the implicit values in science are clarified in the prediction of the future according to science, which I discussed in the section above.

The future of ethical practice through the lens of wisdom looks to the Ultimate Wisdom of God which is both now but 'not yet'. In this way wisdom can challenge those projects which assume that the future is one which just emerges from the present. If science is to have an ethic, a value that is truly rooted in a knowledge of Being, as I argued earlier in this book, then science must learn to listen to the voice of wisdom. I can only give a dim outline of the shape of what this would mean in practical terms. However, one way where this could start would be in the design and implementation of science policy. Those responsible for making decisions for funding need to ask not just: 'Does this fit in with the logic of our current state of knowledge?' or, more insidiously: 'Will this make a profit?', but: 'Is this wise?' 'What are the long-term as well as the short-term benefits?' 'What effect will this have on the social and cultural contexts as well as the environmental context? It is getting a grip on the complexity of each problem that can seem daunting and as a result it is tempting to ignore the complexity in favour of one or other factors.

The pressure to just look at one issue, to the exclusion of all others, is part of scientific method which has conditioned us to think in this way in our ethical decisions as well. As Busch *et al.* have commented:

> It makes little sense to talk of the new biotechnologies independently of the social, political, economic and scientific networks of which they are a part. It is only in this setting, within this complex network of social

> relations in which all actors are involved in continuing negotiation, persuasion, and coercion, that we can talk of impacts.[55]

One of the failures in biotechnology seems to be a failure to address the issue of who takes responsibility. Furthermore, we need to be clear about what these responsibilities are and who takes account of positive responsibilities, that is duties which are specific to this profession.[56] A clarification and demarcation of responsibilities, including protecting the environment, keeping a check on the advantage in the marketplace and consumer interests and concerns, all require proper use of wisdom. Busch *et al.* have suggested that we need adequate assessment of:

1. future health risks associated with biotechnology;
2. future environmental consequences;
3. future burdens and benefits associated with the transformation of institutions;
4. an assessment of decrease in quality of life associated with shift in nature and focus of science resulting from biotechnology research and development.[57]

I would like to add a number five, namely an assessment of the social and cultural impact arising directly out of the new technologies. While it is linked to number four, it puts emphasis on cultural and social impact, rather than 'quality of life', which could imply restriction to material good or restriction to quality as defined from a Western perspective.

Implications of wisdom for 'sustainability'

Earlier in this book I was critical of the way sustainability has been imported by those of apparently very different goals. For environmentalists, sustainability seems to endorse environmental action within an overall development policy. For technocrats, sustainability may be used as a means to justify action, regardless of the impoverishment to life: in essence, as long as resource needs are met it seems we could, for example, replace trees with plastic ones and still adhere to a form of 'sustainability' defined in economic terms. The use of sustainability as a blanket term in order to endorse any and every policy, by changing the meaning of the term itself, seems to me to reflect the profound lack of wisdom in much public discussion of the environment and how to integrate this with development concerns. Other blanket terms which seem to justify actions, regardless of their inherent value, emerge in rhetoric on animal rights, rights of women and so on. It is as if the emotive factor in the language is used as a cover-up for the truth that underlies these statements.

A reconstruction of such language through wisdom would acknowledge the real difference between the kind of sustainability that is advocated by multinational companies that have profitability as the real motivation and the sustainability of local peasant farmers living on the margins of society. Furthermore, wisdom looks for distinctions within multinationals and is able to separate the honest search for corporate responsibility from the need for public approval and the use of ethical terminology simply as a lubricant for society. The public language of sustainability seems to be one which takes as its starting point the reductionist model of science, namely that nature is there as 'object'. This may be one reason why there is public disaffection with the official programmes of sustainability.[58] Szerszynski suggests that it is the language of sustainability which strikes the listener in a hostile way, since its modernist assumptions try to impose an impoverished view of the human person.[59] The modernist view sees the person as an autonomous, bounded entity, pursuing goals through rational calculations about possible outcomes of his or her actions.[60] While there has been a gradual shift in some areas, such as social psychology, to a fuller perception of the human, institutional language remains firmly ingrained in the language of modernism.

There may be other reasons for public disaffection with the language of sustainability; namely that its use has become so variegated that there is confusion about what it means. Adoption of the slogan by so many political parties and interest groups is short-sighted: there is now a paling in not just enthusiasm for a more holistic form of development, but interest in environmental concerns as such. It is as if the language itself has outworn its usefulness. The question now becomes: how can we introduce a language that seems to be resonant with public concern but will have enduring appeal? It seems to me that the language of wisdom is capable of generating a paradigm that will foster positive attitudes to the environment, since it has due regard for history and ancient resourcefulness. It is a language of poetry, drama and spirituality which is rooted in pragmatism and experience. Yet it is also a language which gives hope in the face of the nihilism opened up by the collapse of the modernist ideal of progress. In this sense it is a reassuring term, one which echoes the feminine and gives a richer understanding of humanity as such, as well as in relation to the natural world. The words I have suggested to replace sustainability are 'holistic development'. I use the word holistic, which is part of common currency, as I am cautious about how far it is possible to introduce a brand new word. The word holistic puts emphasis on relationality, on integration of development and environmental needs. Furthermore it is grounded not so much in an ethic of the need to preserve the world for future generations, which is the background for sustainability, but in the here and now: what is possible and feasible for the benefit of all – humankind, the natural world and the world in space.

Environmental philosopher Keekok Lee has suggested that our definition of 'nature' needs to extend to other planets as well.[61] We need to have an astronomically bound ethic, not an earthbound one. In this age of technology, which seems to threaten to colonize other planets, this seems to be a wise shift in emphasis, that is as long as we do not lose sight of our priorities for this planet. Lee proposes that we need to develop both awe and humility in our relationship with other planets. In this scenario it would be morally irresponsible for humans wantonly to destroy a lifeless planet. She suggests that any definition of intrinsic value in terms of interests misses out our extraplanetary responsibilities. Her advocacy of awe and humility seems appropriate for the present discussion, as these are also connected with wisdom; wisdom renders an individual both humble and full of wonder. The loss of the capacity to wonder leads to the impoverishment of the human person and the person in relation to all things, including other planets. Furthermore, while sustainability seems to depend on a narrow definition of the earth, it fails to consider potential exploitation of other planets. However, I am not advocating holism, which is locating value in the whole (even a stellar whole) to the exclusion of the individual. Rather, it is holism as understood through the paradigm of wisdom, which respects both the individual person and the cosmic reality of which we are part. It is a re-rootedness of the individual in the community and in the cosmic.

Conclusions

We are faced today with a plethora of challenges to our human identity, in particular the way we relate to the natural world. The risks to our health through scares such as the BSE crisis; the uncovering of cruelty to animals in the export of live calves in crates for veal production; the poisoning of food through overuse of pesticides; the unease over transplant of live organs from other species to humans as well as transgenetic transformations in the non-human sphere: to name just some examples, these add up to a confusing and somewhat depressing scenario. This threat is a rather different one from the more straightforward prophecy of environmental collapse in the 1960s and 1970s which has become like a shadow in the background of present discussions. The threat to the quality of life is against the background of increasing public disaffection from institutions, be it social, political, ecclesiastical or scientific authority.

The alienation from the environment which has emerged from the modernist attitude of nature as object has now reaped its reward: we seem to be alienated both from ourselves and each other. I suggested that one of the ways through this modern problem is through the generation of new kinds of language; one which draws on the cultural shift to small community action and interaction. The language of wisdom resonates not just with this modern culture, but with the conjunctive language of the ancients. The Celtic saints were among those whose theology and attitude to nature was one that was full of wisdom, in affirmation of all of creation within a framework of self-restraint and positive regard for the Other.

The theme of wisdom encompasses the full range of theological perspectives: from Russian Orthodoxy to radical feminism. Such a re-appropriation of wisdom into our understanding of who God is leads to a rather different perspective on the relation between God, humanity and nature. This different perspective amounts to a New Paradigm, one that echoes the shift towards a new paradigm for research method in being dialectical, experimental and open to the truth.[62] Nonetheless, while wisdom shares some of these qualities of relationality, it allows for content as well as form; the content is expressed in identifying wisdom (Sophia) with *ousia* in God; the form in the emergence of wisdom from practical experience. It thus avoids some of the difficulties of neo-modernism which seem to focus on form at the expense of content, and a fixed traditionalism which ignores form altogether.

What are the possible outcomes of this paradigm on the practice of science and technology and environmental ethics? The future predicted by science seems to be a bleak one, the prophets of artificial intelligence forecast a grim future of highly rational entities that bear no resemblance to the human society as we know it. It seems to be the logical outcome of an impoverished view of the human person as defined by modernism. The public disaffection with such impoverished views seems to me to speak of the wisdom of common sense and intuitive modes of thinking which emerge from the basic experience of life. Instead, an environmental ethic grounded in wisdom is both personal, yet organic, generating hope in the future rather than despair and apathy. Wisdom as integrated into scientific policy would look to the wider social and cultural effects of change, rather than one factor in isolation of all the others. This ability to see the big picture, to look at the whole story, is the capacity which springs from wisdom.

A cultivation of wisdom would also lend itself to consideration of the earth in relation to the other planets. While Gaia concentrates on the interrelatedness on earth, wisdom embraces the cosmos, but still affirms the individual and the human. Instead of sustainability we arrive at holistic development, one which takes into account the social, spiritual and cultural needs of the people in their own context. Wisdom encourages us to think both/and, not either/or. It is both the needs of the people and the needs of the environment that have to be brought together in a balanced relationship. Yet as a relational term, wisdom insists on respect for the Other, rather than mergence with the Other.

Wisdom is not a romantic escape into the medieval era, depending as it does on the authority of given texts as if they had an ontological fixity along with a fixed cosmos. Rather, wisdom incorporates the science of the Enlightenment, but looks to reshape it along different lines. It is a gathering up of what is positive in all our history, rather than ignoring what has taken place. Facing up to the difficulties of the present also means facing up to the negative, a *metanoia* from old attitudes which have been damaging and destructive. It is here that the wisdom of the New Testament speaks volumes. For it allows for suffering and evil by pointing to Christ's death and resurrection as the ultimate source of wisdom. A theological approach to wisdom has no room for arrogance and a false Utopia. Rather, it admits the fact of suffering and evil, but refuses to give them the last word. The God of Wisdom takes on the suffering and evil and challenges all human attempts to find wisdom as if it is a possession to be grasped and used for human aggrandisement. In this way wisdom is not based on an absolute ideal, but it is qualified by the neo-modernist's recognition that all forms of knowledge are imperfect.

My hope is that this book will have in some way contributed to the goal of leading to a shift in human attitudes to the environment, leading to a technology and a science that are more holistic and ethically responsible. Furthermore, I hope that this book has helped to generate a realistic hope that integrates themes from theology, philosophy, technology and science in a new partnership of trust and cooperation. Such an approach can increase our capacity to wonder, to be full of awe at the beauty of the cosmos, in celebration of life as individuals and in community with the natural world.

NOTES

1. R. Grove-White, 'Environment and Society: Some Reflections', *Environmental Politics*, vol. 4, no. 4 (Winter 1995), pp. 265–75.
2. *Ibid.*, p. 271.

3. B. Szerszynski, 'On Knowing What to Do: Environmentalism and the Modern Problematic' in *Risk, Environment and Modernity: Towards a New Ecology*, S. Lash, D. Szerszynski and B. Wynne (eds) (Sage Publications, London, 1996), pp. 104–37.
4. *Ibid.*, p. 109.
5. *Ibid.*, p. 111.
6. J. S. Dyysek, 'Ecology and Discursive Democracy: Beyond Liberal Capitalism and the Administrative State', *Capitalism, Nature, Socialism*, vol. 3, no. 2 (1992), pp. 18–42.
7. J. Habermas, 'New Social Movements', *Telos*, vol. 49 (1981), pp. 33–7.
8. B. Szerszynski, *op. cit.*, p. 118.
9. *Ibid.*, pp. 122–3.
10. M. Heidegger, *Being and Time*, trans. J. MacQuarrie and E. Robinson (Blackwell, Oxford, 1978).
11. D. M. Fiere, 'What is the Appearance of Divine Sophia?', *Slavic Review*, vol. 48, no.3 (1988), pp. 449–76. I have written an expanded version of this section (p. 137) and the following two sections (pp. 139 and 143) in an article entitled 'Sophia: The Feminine Face of God as a Metaphor for an Ecotheology', *Feminist Theology*, 1997, in press.
12. C. H. Chen, *Sophia: The Science Aristotle Sought* (Gerg Olms Verlag, Hildeshem/New York, 1976), p. 384.
13. After Clement of Alexandria, in Greek theology wisdom is more frequently identified with the Son or Logos, which reached its climax in gnostic speculation. Rees believes that Greek theology became Dualitarian, rather than Trinitarian such that: 'It thinks only in terms of God and his Logos; and if the tradition and experience of the Holy Spirit still claim recognition, the system can only admit it as a shadowy repetition of the Logos, with no independent and effective function or principle of its own.' T. Rees, 'The Holy Spirit as Wisdom', in *Mansfield College Essays*, A. M. Fairbairn (ed.) (Hodder and Stoughton, London, 1908), p. 304.
14. C. Bamford, foreword to S. Bulgakov, *Sophia: The Wisdom of God: An Outline of Sophiology* (Lindisfarne Press, Hudson, 1993), p. xvi.
15. S. Bulgakov, *ibid.*, p. 21.
16. *Ibid.*, p. 49.
17. His notion of divine and creaturely Sophia led to the charge of Platonism. Vladimir Lossky believes that Bulgakov has put the personhood of God before God's nature, since wisdom is the common revelation of the persons. He considers that this has come out of an exaggerated reaction against the Western theological tradition which tends to view the Holy Spirit as no more than the bond between Father and Son. The energies of God manifest the wisdom, life, power and justice of God, in other words wisdom is just one of the energies of God. See V. L. Lossky, *The Mystical Theology of the Eastern Church*, trans. the Fellowship of St Alban and St Sergius (J. Clarke, London, 1957), pp. 79–80.
18. B. Newman, 'Sergius Bulgakov and the Theology of Divine Wisdom', *St Vladimir's Theological Quarterly*, vol. 22 (1978), pp. 39–73.
19. J. Zizioulas, *Being as Communion* (St Vladimir's Seminary Press, Crestwood, 1993).
20. C. Graves, *The Holy Spirit in the Theology of Sergius Bulgakov* (World Council of Churches, Geneva, 1972), p. 26.
21. C. Keller, 'Towards an Emancipatory Wisdom', in *Theology and the University: Essays in Honor of John B. Cobb, Jr*, D. R. Griffin and J. C. Hough (eds) (State University of New York Press, Albany, 1991), pp. 125–47.
22. *Ibid.*, p. 143. See also E. Moltmann-Wendel, 'Self-Love and Self-Acceptance', *Pacifica*, vol. 5 (1992), pp. 288–300.
23. E. A. Johnson, *She Who Is: The Mystery of God in Feminine Theological Discourse* (Crossroad, New York, 1992).
24. *Ibid.*, p. 50.
25. *Ibid.*, p. 54. Italics mine.
26. *Ibid.*, p. 169. For a fuller discussion of the relationship between wisdom and Christology, see M. Scott, *Sophia and the Johannine Jesus* (Supplement Series No. 71, Journal for the Study of the New Testament (JSOT) The Almond Press, Sheffield, 1992); E. Schüssler Fiorenza, *Jesus: Miriam's Child, Sophia's Prophet* (SCM Press, London, 1995).
27. E. A. Johnson, *op. cit.*, p. 179.
28. C. V. Camp, *Wisdom and the Feminine in the Book of Proverbs* (JSOT, The Almond Press, Sheffield, 1985).

29. *Ibid.*, p. 277.
30. K. J. Dell, '"Green" Ideas in the Wisdom Tradition', *Scottish Journal of Theology*, vol. 47, no. 4 (1994), pp. 423–51.
31. M. F. Johnson, 'The Sapiential Character of the First Article of the *Summa Theologiae*' in *Philosophy and the God of Abraham: Essays in Memory of James A. Weisheipl*, R. J. Long (ed.) (Pontifical Institute of Medieval Studies, Toronto, 1991), essay pp. 85–98; quotation p. 89.
32. See, for example, Rosemary Radford Ruether, *Gaia and God* (SCM Press, London, 1992).
33. I am, in general, critical of attempts to create a theology from Lovelock's thesis; see C. Deane-Drummond, 'God and Gaia: Myth or Reality?', *Theology* (July/August, 1992), pp. 275–83; C. Deane-Drummond, *A Handbook in Theology and Ecology* (SCM Press, London, 1996), pp. 98–114.
34. I am aware that some feminist theologians have used Sophia to highlight the idea of the goddess. However, it is my intention to link Sophia with more traditional images, rather than use it as a basis for goddess theology. See, for example, S. Cady, M. Ronan and H. Taussig, *Wisdom's Feast: Sophia in Study and Celebration* (Harper and Row, San Francisco, 1989). For a critical commentary of their work, see B. Witherington III, 'Three Modern Faces of Wisdom', *Ashland Theological Journal*, vol. 25 (1993), pp. 96–122. Caitlin Matthews, similarly, identifies Sophia as the Goddess and discusses her influence in the religious consciousness of Western Europe. See C. Matthews, *Sophia; Goddess of Wisdom: The Divine Feminine from Black Goddess to World Soul* (Mandala/Harper Collins, London, 1991).
35. For a discussion, see H. J. Hermisson, 'Observations on the Creation Theology in Wisdom', in *Israelite Wisdom: Theological and Literary Essays in Honour of Samuel Terrien*, J. G. Gummie, W. A. Brueggeman, W. L. Humphreys, J. M. Ward (eds) (Scholars Press, Missoula, 1978), pp. 43–57.
36. See, for example, J. Clague, 'Interview with Margaret Argyle', *Feminist Theology*, vol. 10 (1995), pp. 57–68.
37. J. Moltmann, 'The Motherly Father: Is Trinitarian Patripassionism Replacing Theological Patriarchalism?', trans. G. Knowles, in *God as Father? Concilium 143*, J. Metz and E. Schillebeeckx (eds) (T. and T. Clark, Edinburgh, 1981), pp. 51–6.
38. F. Watts, 'Science and Eschatology', *Modern Believing*, vol. 36, no. 4 (October 1995), pp. 46–52
39. For further discussion, see C. Deane-Drummond, 'Jürgen Moltmann on Heaven', in *Spirit World*, A. Lane (ed.) (Paternoster Press, Carlisle, 1996), pp. 49–64.
40. For a commentary on Moltmann's dialogue with secular notions of hope in his earlier work, such as *Theology of Hope* (SCM Press, London, 1967), see R. Bauckham, *The Theology of Jürgen Moltmann* (T. and T. Clark, Edinburgh, 1995), pp. 43–6.
41. E. Bloch, *Principle of Hope* (Blackwell, Oxford, 1986), pp. 1372ff. For a discussion of Moltmann's interpretation of Bloch, see J. Moltmann, *God in Creation* (SCM Press, London, 1985), pp. 178–91.
42. J. Moltmann, *God in Creation*, *ibid.*, p. 179.
43. E. Bloch, *op. cit.*, p. 1373.
44. F. Watts, *op. cit.*, pp. 47–50.
45. C. J. S. Clarke, 'Review of F. Tipler, *The Physics of Immortality*' (Macmillan, London, 1995), *Theology in Green* (Summer 1995), pp. 45–7. Italics mine.
46. Richard Dawkins's classic text is *The Selfish Gene* (Oxford University Press, Oxford, 1976).
47. D. King, 'Ill Fitting Genes', *The Times Higher Educational Supplement* (14 June 1996), p. 19.
48. M. Poole, 'A Response to Dawkins', *Science and Christian Belief*, vol. 7, no. 1 (April 1995), pp. 51–8.
49. E. Sahouris, *The Human Journey from Chaos to Cosmos* (Pocket Books, London, 1989).
50. For further discussion, see C. Deane-Drummond, 'God and Gaia: Myth or Reality?', *op. cit.*, pp. 275–83 and C. Deane-Drummond, 'Gaia as Science Made Myth: Implications for Environmental Ethics', *Studies in Christian Ethics*, vol. 9, no. 2 (1996), pp. 1–15.
51. F. Ferré, 'Personalistic Organicism: Paradox or Paradigm?', in *Philosophy and the Natural Environment*, R. Attfield and A. Belsey (eds) (Cambridge University Press, Cambridge, 1994), p. 59.

52. *Ibid.*, p. 70.
53. R. Grove White, 'Human Identity and the Environmental Crisis', in *The Earth Beneath*, I. Ball, M. Goodall, C. Palmer and J. Reader (eds) (SPCK, London, 1992), pp. 13–34.
54. H. Küng, *Global Responsibility: In Search of a New World Ethic* (SCM Press, London, 1992).
55. L. Busch, W. B. Lacy, J. Burkhardt and L. R. Lacy, *Plants, Power and Profit: Social, Economic and Ethical Consequences of the New Biotechnologies* (Blackwell, Oxford, 1992), pp. 191–2.
56. *Ibid.*, pp. 204–5.
57. *Ibid.*, p. 212.
58. R. Grove-White, 'Environmental Knowledge and Public Policy Needs: On Humanising the Research Agenda', in S. Lash, D. Szerszynski and B. Wynne (eds), *op. cit.*, pp. 271–85.
59. B. Szerszynski, 'Sustainable Development and Human Identity', unpublished communication.
60. E. E. Sampson, 'The Deconstruction of the Self', in *Texts of Identity*, J. Shotter and K. E. Gergen (eds) (Sage Publications, London, 1989), pp. 1–19.
61. K. Lee, 'Awe and Humility: Beyond an Earthbound Environmental Ethics', in R. Attfield and A. Belsey (eds), *op. cit.*, pp. 89–101.
62. P. Reason and J. Rowan (eds), *A Dialectical Paradigm for Research* (Wiley, London, 1981).

Index